Designing Inclusive Futures

Patrick Langdon • John Clarkson • Peter Robinson
Editors

Designing Inclusive Futures

 Springer

Editors

Patrick Langdon, BSc PhD
John Clarkson, MA, PhD, CEng, MIEE

Cambridge Engineering Design Centre
Department of Engineering
University of Cambridge
Trumpington Street
Cambridge
CB2 1PZ
UK

Peter Robinson, MA, PhD, Ceng

Computer Laboratory
University of Cambridge
William Gates Building
15 JJ Thomson Avenue
Madingley Road
Cambridge
CB3 0FD
UK

ISBN 978-1-84800-210-4   e-ISBN 978-1-84800-211-1

DOI 10.1007/978-1-84800-211-1

British Library Cataloguing in Publication Data
A catalogue record for this book is available from the British Library

Library of Congress Control Number: 2008921347

Printed on acid-free paper

9 8 7 6 5 4 3 2 1

springer.com

Preface

The Cambridge Workshop on Universal Access and Assistive Technology (CWUAAT) 2008 is the fourth of a series of workshops that are held every two years. The workshop theme, *Designing inclusive futures*, reflects the need to explore in a coherent way the issues and practicalities that lie behind design intended to extend our active future lives. This encompasses design for inclusion in daily life at home but also extends to the workplace and to products within these contexts. For example, given trends in employment sector growth, skills requirements, labour supply and demographic change, there is a need to predict the critical areas where individual capabilities are mismatched with the physical, social and organisational demands of work. This mismatch, which can be addressed within the domain of inclusive design, is pervasively linked to real artefacts in workspaces and their intersection with the health factors that relate to ageing.

Since the inception of the current CWUAAT series in 2002, the scope and specific areas of focus of the workshop have changed, reflecting new directions in both the research and social context. Hence, we have moved from isolated activities in disparate fields such as engineering, therapy and computer science, that were reflected in the first volume's title, *Universal access and assistive technology*, via the integrating themes of *Designing a more inclusive world* and *Designing accessible technology*, to the more unified and holistic perspective evident in the present title: *Designing inclusive futures*. In the context of inclusive design, this reflects the development of theories, tools and techniques as research moves on, and also the need to draw in wider psychological, social, and economic considerations in order to gain a more accurate understanding of user interactions with products and technology. Two consequences of this are evident in the book content: firstly, the increased emphasis on the transfer of knowledge and techniques from research into the wider community as research programs mature, and secondly, the requirement for connected thinking that can encompass multi-product sequences of actions that are a part of tasks or workplace activities in a variety of social settings.

The book contains the best papers submitted for CWUAAT'08, as judged by an international panel of thirty three reviewers. This panel and the final contributors represent a sample of leading national and international researchers in the fields of inclusive design, ergonomics, universal access, and assistive and rehabilitative technology. We are also pleased to note that there have been significant contributions from researchers in architecture, social housing provision and apparel and fashion design, reflecting the need to understand the wider social and economic context of inclusive and assistive technology design.

This book is divided into the following five parts:

I.   *Understanding Users* focuses on the identification and provision of usable data about user capabilities to designers and methods of using it to reduce exclusion.

II.   *Inclusive Design* presents case studies of inclusive design and research on the theory of applying it to products and daily living.

III. *Computer Access and New Technologies* looks at research into ICT accessibility and new technology for inclusion, demonstrating that these are still priority focus areas.

IV.   *Assistive Technology* focuses on special purpose design and adaptations for specific impairments.

V.   *Inclusive Environments* examines the context of the inclusive design of products and encompasses architecture, usability, housing requirements and ageing.

As is usual for CWUAAT, the nature of the contributions has been broad, both within individual themes and also across the workshop's scope. We hope this will lead to new solutions to reduce exclusion and difficulty arising from impairment, with special application to our future lives, in the workplace, at home and at leisure.

As ever, we would like to thank all those who contributed to the CWUAAT'08 Workshop and to the preparation of this book. Many thanks are due to all the contributors and also to the non-contributing members of the Programme Committee for the high quality of the reviewing. Thanks are particularly due to Mari Huhtala, who, together with Suzanne Williams, now plays a far greater role in bringing the resulting publication to fruition than ever before. We would also like to thank the staff at Fitzwilliam College who enable us to run the workshop.

Pat Langdon, John Clarkson and Peter Robinson
The CWUAAT Editorial Committee
University of Cambridge
April 2008

Contents

Part II Inclusive Design

Part III Computer Access and New Technologies

Part IV Assistive Technology

Part V Inclusive Environments

List of Contributors

Afacan Y., Department of Electrical and Electronics Engineering, Bilkent University, Bilkent, Ankara, Turkey

Andrews B.J., School of Technology, Oxford Brookes University and Nuffield Department of Surgery, Oxford, UK

Baker P.M.A., Wireless RERC, Georgia Institute of Technology, Atlanta, GA, US

Baskinger M., School of Design, Carnegie Mellon University, Pittsburgh, PA, US

Bates R., De Montfort University, The Gateway, Leicester, UK

Biswas P., Computer Laboratory, University of Cambridge, Cambridge, UK

Bix L., School of Packaging, Michigan State University, MI, US

Blessing L., Engineering Research Unit, University of Luxembourg, Luxembourg

Bowen S.J., Art and Design Research Centre, Sheffield Hallam University, Sheffield, UK

Bush T.R., Mechanical Engineering, Michigan State University, MI, US

Carroll K.E., Department of Interior Design and Merchandising, East Carolina University, Greenville, NC, US

Chadd J., Bath Institute for Medical Engineering, School for Health, University of Bath, Bath, UK

Chalfont G., School of Architecture, University of Sheffield, Sheffield, UK

Chamberlain P.M., Art and Design Research Centre, Sheffield Hallam University, Sheffield, UK

Choi Y.M., College of Architecture, Industrial Design, Center for Assistive Technology and Environmental Access, Georgia Institute of Technology, Atlanta, GA, US

Clarke S., Department of Civil and Environmental Engineering, Imperial College London, London, UK

Clarkson P.J., Engineering Design Centre, Department of Engineering, University of Cambridge, Cambridge, UK

Dong H., School of Engineering and Design, Brunel University, Uxbridge, Middlesex, UK

Fair J.R., School of Packaging, Michigan State University, MI, US

Feng J., Department of Computer and Information Sciences, Universal Usability Laboratory, Towson University, Towson, MD, US

Froyen H., Department of Architecture, PHL, Hasselt University, Belgium; and Faculty of Engineering, Ghent University, Gent, Belgium

Gibbons R., ASPIRE National Training Centre, Stanmore, London, UK

Gibson G., Department of Primary Care, University of Liverpool, Liverpool, UK

Gooch S., Mechanical Engineering, University of Canterbury, Christchurch, New Zealand

Goodey S., London Regatta Centre, London, UK

Goodman-Deane J., Engineering Design Centre, Department of Engineering, Cambridge University, Cambridge, UK

Hanington B., School of Design, Carnegie Mellon University, Pittsburgh, PA, US

Hanson J., Bartlett School of Graduate Studies, University College London, London, UK

Hettinga D., School of Health and Social Care, Brunel University, London, UK

Heylighen A., Department of Architecture, Urbanism and Planning, Katholieke Universiteit Leuven, Leuven, (Heverlee), Belgium

Hicks B.J., IdMRC, Department of Mechanical Engineering, University of Bath, Bath, UK

Holman J., Department of Computer and Information Sciences, Universal Usability Laboratory, Towson University, Towson, MD, US

Howcroft D., Research Institute for the Care of Older People, Bath, UK

Hughes G., Computer Laboratory, University of Cambridge, Cambridge, UK

Hurtienne J., Center of Human-Machine-Systems and Department of Mechanical Engineering and Transport Systems, TU Berlin, Berlin, Germany

Istance H.O., De Montfort University, The Gateway, Leicester, UK

Janson R., Engineered Packaging, Department of Mechanical Engineering, University of Sheffield, Sheffield UK

Kincade D.H., Department of Apparel, Housing, and Resource Management, Virginia Tech, Blacksburg, VA, US

Kwok Y.C.J., School of Design, Hong Kong Polytechnic University, Kowloon, Hong Kong

Langdon P.M., Engineering Design Centre, Department of Engineering, University of Cambridge, Cambridge, UK

Langley J., Human Centred Engineering, Art and Design Research Centre, Sheffield Hallam University, Sheffield, UK

Lazar J., Department of Computer and Information Sciences, Universal Usability Laboratory, Towson University, Towson, MD, US

Lewis T., Engineering Design Centre, Department of Engineering, Cambridge University, Cambridge, UK

Luxmoore J., Engineered Packaging, Department of Mechanical Engineering, University of Sheffield, Sheffield UK

Mayagoitia R.E., Centre of Rehabilitation Engineering, King's College London, London, UK

Medland A.J., IdMRC, Department of Mechanical Engineering, University of Bath, Bath, UK

Moon N.W., Wireless RERC, Georgia Institute of Technology, Atlanta, GA, US

Ng C.H.K., Department of Architecture, University of Hong Kong, Hong Kong

Orpwood R., Bath Institute for Medical Engineering, School for Health, University of Bath, Bath, UK

Poulton A., Department of ICT, The Open University, Milton Keynes, UK

Robinson P., Computer Laboratory, University of Cambridge, Cambridge, UK

Rowson J., Engineered Packaging, Department of Mechanical Engineering, University of Sheffield, Sheffield UK

Rychtáriková M., Department of Physics and Astronomy, Katholieke Universiteit Leuven, Leuven, (Heverlee), Belgium

Sabata D., Center for Assistive Technology and Environmental Access, Georgia Institute of Technology, Atlanta, GA, US

Sixsmith A., Department of Primary Care, University of Liverpool, Liverpool, UK

Sprigle S., Center for Assistive Technology and Environmental Access, Georgia Institute of Technology, Atlanta, GA, US

Taylor J.C., Engineered Packaging, Department of Mechanical Engineering, University of Sheffield, Sheffield UK

Tinker A., Institute of Gerontology, King's College London, London, UK

Todd R., Center for Assistive Technology and Environmental Access, Georgia Institute of Technology, Atlanta, GA, US

Torrington J., School of Architecture, University of Sheffield, Sheffield, UK

van Boxstael E., Centre of Rehabilitation Engineering, King's College London, London, UK

Vanns N., Nokia (UK) Ltd., Nokia House, Farnborough, UK

Vermeir G., Department of Civil Engineering, Katholieke Universiteit Leuven, Leuven (Heverlee), Belgium

Vickers S., De Montfort University, The Gateway, Leicester, UK

Waller S.D. Engineering Design Centre, Department of Engineering, Cambridge University, Cambridge, UK

Weber K., Department of Economics I, University of Applied Sciences (FHTW) Berlin, Germany

Wojgani H., Bartlett School of Graduate Studies, University College London, London, UK

Wright F., Institute of Gerontology, King's College London, London, UK

Yoxall A., Human Centred Engineering, Art and Design Research Centre, Sheffield Hallam University, Sheffield, UK

Part I

Understanding Users

Chapter 1

Converting Disability Data into a Format Suitable for Estimating Design Exclusion

S.D. Waller, P.M. Langdon and P.J. Clarkson

1.1 Introduction

The Cambridge Engineering Design Centre is unique in developing analytical tools that can quantitatively assess the inclusive merit of different design decisions, according to the number of potential users that would be excluded: such tools would greatly assist the implementation of inclusive design in businesses (Dong *et al.*, 2003). In addition to those excluded from using a product, many more people will experience difficulty or frustration, so reducing the number of people excluded can improve the experience for a wide range of users. Indeed research commissioned by Microsoft (2003) reported that "60% of Americans aged 18-64 years were likely or very likely to benefit from the use of accessible technology".

Exclusion auditing is intended to complement other tools for evaluating inclusive merit, such as expert opinion (Poulson *et al.*, 1996), user trials (Aldersey-Williams *et al.*, 1999), and impairment simulators (Steinfeld and Steinfeld, 2001). In combination, these tools provide a holistic approach to discovering the causes of design exclusion, and identifying appropriate design improvements.

A useful data source for evaluating design exclusion should be generalisable to the UK national population, and have sufficient scope to cover all of the capabilities involved in product interaction, so that issues with double counting can be resolved (Keates and Clarkson, 2003). The data must also be concise enough to avoid data overload during its use, and simple to apply for the evaluation of product interaction.

The 1996/97 disability follow-up survey (Grundy *et al.*, 1999) remains the most recent dataset that satisfies all of these criteria. It contains data regarding the proportion of the Great Britain (GB) population who can or cannot perform everyday tasks, grouped into categories that directly relate to the cycles of perceiving, thinking and acting that occur during product interaction. The methodology behind this survey is first explored, followed by an explanation of how the survey data is converted into graphs that measure demand and exclusion for a product, and guidance on how these graphs can be used.

1.2 Introduction to the Disability Follow-up Survey

The DFS was performed to measure the prevalence of disability within Great Britain, in order to help plan welfare support. Disability was defined as "any restriction or lack of ability to perform an activity in the manner or within the range considered normal for a human being" according to the World Health Organisation (WHO, 1980) International Classification of Impairments Disabilities and Handicaps. Adults were selected for the DFS if they met certain criteria, such as being in receipt of incapacity benefit; those under the age of 16 were not included. Approximately 7,500 participants were asked up to 300 questions regarding whether they were able to perform certain tasks such as:

- "Can you pick up a safety pin with your left hand?"
- "Can you tie a bow in laces or string without difficulty?"

The questions were grouped together in 13 ability categories, seven of which are most relevant for product interaction, namely seeing, hearing, intellectual function, communication, locomotion, reach and stretch, and dexterity (seeing and intellectual function are henceforth renamed vision and thinking for clarity). Within each ability category, a panel of approximately 100 judges collated the questions to form a set of ability levels. The judges included a range of professionals with experience of disability, independent researchers working in the field, staff involved with the survey, and disabled people and their carers (Martin and Elliot, 1992).

The statements used to describe the dexterity ability levels are shown in Figure 1.1, together with the proportion of the GB adult population within each level. The dexterity ability levels range from extremely low ability, such as level D1 "Cannot pick up and hold a mug of coffee with either hand" to moderate ability, such as level D8 "Has difficulty wringing out light washing". Level D12 represents full dexterity ability.

The complete set of statements used to describe all the ability levels in each of the seven relevant categories were first published in Martin et al. (1988), and can also be found in Keates and Clarkson (2003). The language used to describe the ability levels takes several different forms, such as "Cannot do something", "Has difficulty doing something", or "Can do something but not something".

Inconsistencies in language use, and the sheer number of different ability levels make it difficult to quickly discover the underlying functions that are covered within each of the scales, and how these functions vary as the scale progresses. Experience gained through performing exclusion audits and providing inclusive design training identified a method to resolve these issues, which is now presented. The data is converted into a set of graphs that directly plot the demands made by a product against the number of people who would be unable to use it. These graphs are intended to help evaluate alternative products, and identify the causes of design exclusion.

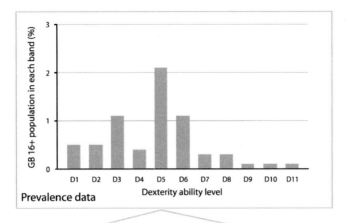

D1   Cannot pick up and hold a mug of coffee with either hand
D2   Cannot turn a tap or control knob on a cooker with either hand
D3   Cannot pick up and carry a pint of milk or squeeze the water from a
        sponge with either hand
D4   Cannot pick up a small object such as a safety pin with either hand
D5   Has difficulty picking up and pouring from a full kettle, or serving food
        from a pan using a spoon or ladle
D6   Has difficulty unscrewing the lid of a coffee jar, or using a pen
D7   Cannot pick up and carry a 5lb bag of potatoes with either hand
D8   Has difficulty wringing out light washing or using a pair of scissors
D9   Can pick up and hold a mug of tea with one hand but not the other
D10 Can turn a tap or control knob with one hand but not the other. Can
        squeeze water from a sponge with one hand but not the other
D11 Can pick up a small object such as a safety pin with one hand but not
        the other. Can pick up and carry a pint of milk with one hand but not
        the other. Has difficulty tying a bow in laces
D12 Full dexterity ability

Dexterity ability levels

Figure 1.1. Prevalence data for dexterity from the disability follow up survey

1.3 Constructing Graphs of Demand and Exclusion

Interacting with a product will typically place demands on up to seven of the capability categories identified within the DFS. In order to produce graphs for demand and exclusion for each of these categories, the demands associated with product interaction that are covered within the DFS data are first identified, and simple models are presented to explain how this demand varies (according to the underlying ability data). Each demand scale is then characterised by a set of anchor points, created from the original DFS ability statements, then the numerical analysis required to produce the exclusion graphs is presented.

The original scales were developed by health practitioners and therapists to measure the quality of life impairment associated with being unable to perform activities of daily living. As such, they were not originally intended to cover all aspects of user capabilities involved with product interaction. Issues are therefore highlighted that are relevant for product interaction, but were not covered within the DFS data, together with aspects of the context of use that may affect capability demands.

1.3.1 Demands Associated with Product Interaction

For each of the seven capability categories in turn, the demands associated with product interaction that are covered within the DFS data are stated, along with other issues pertinent to evaluating capability demand.

Vision
The DFS data caters for demands where the user is required to do things like:

- reading text of various sizes;
- recognising a friend at various distances.

Assuming that the eye can correctly focus given the viewing distance, a simple vision demand is proportional to the feature size, the distance away, and the brightness and colour contrast between foreground and background.

The context of use also affects the demand level, for example if the ambient lighting is too low, or indeed too high. Problems may also arise according to the angle of view. The DFS did not cover colour blindness, or the usable visual field, so other data sets have to be used to estimate exclusion arising from these factors.

Hearing
The DFS data caters for demands where the user is required to do things like:

- understanding speech against various levels of background noise;
- detecting sounds of various volumes.

A simple hearing demand is proportional to the ratio between the sound to be detected, and the level of ambient noise. The ambient noise level depends on the context of use, and is particularly problematical for those using hearing aids, unless induction (T-coil) loops transmit a signal directly to the hearing aid.

Thinking
The DFS data caters for demands where the user is required do things like:

- understanding or express written language;
- recalling things from memory, or remember to do things in the future;
- holding attention, and process information with clarity.

The judges' first attempt to rank the DFS thinking questions in order of quality of life impairment was rejected, because the variance between the rankings of different judges failed statistical validity checks (Martin and Elliot, 1992). To resolve this issue, a person's ability was calculated by counting up the number of thinking tasks they were able to perform. Correspondingly, the simple thinking demand is proportional to the number of tasks that are similar to those used within the DFS.

A better model for estimating the thinking demand associated with product interaction would consider the demand levels in categories such as perception and response processing, working memory, recognition and recall, attention, visual-spatial thinking, and verbal thinking. An alternative cognitive scale has been constructed from the original DFS questions (Langdon *et al.*, 2003), but thinking demands can only be incorporated with the other six categories if the methodology behind the original thinking scale is preserved.

Communication

The DFS data caters for demands where the user is required to do things like:

- communicating with people who are well-known to the user;
- communicating with strangers.

The level of communication demand depends upon whether the other person is well known to the user, and if the communication must occur "easily", or could occur "albeit with some difficulty". Whether the user is willing to tolerate such difficulty is assumed to depend upon the frequency with which the demand occurs.

The communication scale specifically focuses on the ability to communicate with other people through speech. Even with such a limited definition, a communication demand is coupled together with vision, hearing, and thinking demands; the ability to control the muscles involved with speech generation is also vital. The DFS did not cover which language is first spoken, or cover communication difficulties associated with regional dialects, or other nonverbal aspects of communication.

Locomotion

The DFS data caters for demands where the user is required to do things like:

- walking various distances on level ground;
- ascending or descend stairs;
- balancing without holding on to something;
- bending down to reach something.

A simple walking demand is proportional to the distance required without stopping. A simple climbing demand is proportional to the number of steps, together with the availability of a handrail and resting opportunities. A simple balance demand is proportional to the time that the user has to stand without being able to hold onto something. A simple bending demand is proportional to the distance that the user must reach below waist level, and the amount of time that the user must spend in this position.

The context of use also affects the demand level, according to difficulties with factors such as slope, ground surface, stamina and fatigue, and whether the user needs to carry heavy items while performing the task.

Reach and Stretch
The DFS data caters for demands where the user is required to do things like:

- reaching out in front, up to the head, or behind the back with one or both arms.

The level of reach and stretch demand depends upon the number of arms that need to be extended, and if the user needs to perform the task "easily", or if it could be done "albeit with some difficulty". Whether the user is willing to tolerate such difficulty is assumed to depend upon the frequency with which the demand occurs, or the amount of time that the arm has to be held out for. The context of use also affects the demand level, for example if user is wearing heavy clothing, or needs to carry other items while performing the task.

Dexterity
The DFS data caters for demands where the user is required to do things like:

- performing fine finger manipulation;
- grasping objects;
- picking up and carry objects.

A simple dexterity demand is proportional to the number of hands needed to perform the task, together with the required levels of precision and magnitudes of force. Such variation is elaborated according to different example tasks that need to be performed. For example, the level of precision varies between that required to pick up a safety pin, and that required to tie a bow in shoelaces. The level of force varies between that required to pick up a pint of milk, and that required to pick up a bag of potatoes.

The context of use also affects the demand level, for example if the hands are sweaty or wet, if gloves are worn, and if the ambient temperature is extreme. Vibration, motion and the visibility of the hands can also affect the ability to grip or make precision movements.

1.3.2 Producing Objective Demand Scales

When attempting to evaluate a product according to its demand on each different capability category, the first task is to quickly identify which of the categories are relevant for the interaction. To help this task, a "No demand" statement was written for each of the seven categories, based on the user not being required to perform any of the activities covered within any of the underlying ability levels of that category. For example, the "No demand" statement for dexterity is "The user is not required to pick up and carry objects, or to perform fine finger manipulation, or to grasp objects". In this case no one would be excluded from using the product by its dexterity demand.

For each category that is deemed relevant, the next task is to identify whether the demand level exceeds that covered by the DFS data, so a "High demand" statement was written based on the user being required to perform the hardest tasks contained within that category (other data sources should be consulted if "High demand" is exceeded). Assuming that the task being assessed is covered by the DFS data, then a psychological scaling judgement (Gulliksen and Messick, 1960) is needed to position the demand level somewhere between "No demand" and "High demand". Defining two intermediary anchor points was considered to provide the best way to enhance the precision and reliability of this scaling judgement, without 'data overloading' the assessor. These two intermediary anchor points were named "Low demand" and "Moderate demand".

Continuing the example for dexterity, the "Low demand" statement was constructed to exclude those in ability levels D1-D4, while "Moderate demand" would exclude those in D1-D8. All demands are expressed in a unified language in the form of "The user is required to have sufficient ability to do things like". The resulting demand scale for dexterity is shown in Figure 1.2 on page 10, along with the statements used to define each anchor point.

The four anchor points break the demand scale up into three regions, and the scales are only intended to be accurate to a level where an assessor can quickly identify the correct region, and then roughly decide whether the demand is closer to the middle of the region, or one of the anchor points. This process breaks the demand scale up into approximately nine regions, which broadly matches the resolution of the original data points.

1.3.3 Producing Exclusion Graphs

The prevalence data shown in Figure 1.1 is in ordinal form, because it refers to the number of people that exist in certain categories, and each category exists in a ranked order (D1, D2, D3, *etc.*). The method converting these data points into a format suitable for assessing demand and exclusion is illustrated for the dexterity example in Table 1.1.

Table 1.1. Data points used to construct Figure 1.2.

x-value	x-label	y-value
0	No demand	0
4	Low demand	No. people in D1-D4
8	Moderate demand	No. people in D1-D8
11	High demand	No. people in D1-D11

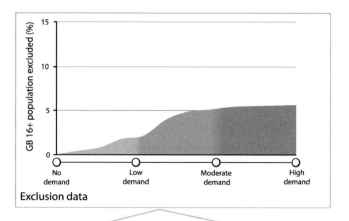

Figure 1.2. Example graph of demand and exclusion for dexterity

The data points that sit in between each anchor point were calculated in a similar manner, although these points were not labelled on the x-axis. The data was made continuous by fitting a cubic spline through all the points, and the resulting graph is shown in Figure 1.2. The demand statements and exclusion graphs for all of the capability categories are presented in Clarkson *et al.* (2007), together with a design example.

1.4 Using the Graphs to Measure Exclusion

A level of demand corresponds with a particular task, with a particular product (or products), in a particular environment. A level of exclusion corresponds with the proportion of people unable to perform that task. The estimated levels of demand and exclusion therefore depend on the assumptions made regarding the task being performed, the current state of any other items relevant to the interaction, and any pertinent features of the environment. The population statistics also assume that any desired vision, hearing or locomotion aids can be used, so the results may need to be modified if the context of use prevents this.

Once the demand level has been identified in each of the seven capability categories, it is often desirable to estimate the overall exclusion caused by the combination of capability demands. Many people have losses in more than one capability category, so simply adding up the number of people excluded due to each individual category would significantly overestimate the overall exclusion. This problem with double counting can easily be overcome using software that is freely available on www.inclusivedesigntoolkit.com. The software uses the DFS results for each of the 7,500 participants, and strikes out each participant if any of the demands exceed their capabilities. National exclusion figures can then be calculated by pro-rating the number excluded to the total GB population.

1.4.1 Further Considerations

Once the overall levels of demand and exclusion associated with a single task have been identified, it may then be of interest to investigate how these combine together over a series of tasks required for product use, so that design improvements can be targeted towards the tasks that cause problems for the greatest numbers of people. If the population that cannot use a product is known, it would also be desirable to calculate how many of these people actually want to use the product. A complete consideration of this matter is a subject for further research, although in some cases it can be estimated using existing capability data. For example, those who cannot drive are unlikely to want to use a satellite navigation system for a car, and those who cannot pick up and drink hot liquid from a mug are unlikely to want to use a kettle. Keates and Clarkson (2003) consider various alternative measures for the inclusive merit of a product, according to the proportion of the market that can use it.

1.5 Conclusions and Further Work

In combination with other tools, an exclusion audit can help to identify the causes of design exclusion, and to prioritise design improvements to reduce this exclusion, thereby improving the product experience for a broad range of users. Estimating exclusion requires a single dataset that covers all of the capabilities required during the cycles of perceiving, thinking, and acting that occur during product interaction. The 1996/97 disability follow-up survey remains the most recent dataset with sufficient scope, although the results from this survey were originally presented in a format suitable for calculating the number of people with various levels of quality of life impairment.

A set of scales that measure the capability demands that occur during product interaction have been developed, which are directly compatible with the original survey data. The authors are now conducting the next phase of research, which will measure the reliability and validity of these scales for assessing design exclusion. However, the maximum achievable reliability and validity is limited, because the underlying data were not collected for this purpose. The statements that define each demand scale therefore unavoidably confound and confuse many of the underlying user capabilities that are required for product interaction. For example the scale that claims to measure vision demand does so according to the ability to read text or recognise a friend, both of which also imply a certain thinking demand.

It may be possible to derive better data from the original answers to the 300 DFS questions, rather than using the ability levels from Martin *et al.* (1988). Ultimately though, quantifying design exclusion requires a new national survey conducted specifically for this purpose, which covers sensory, cognitive, physical, and social causes of exclusion. Designing and undertaking such a survey is one objective of the i~design 3 research project, due for completion around 2009.

Estimating the number of people who might be frustrated or have difficulty with a product would also significantly enhance an exclusion audit. Calculating the exclusion associated with a task that can be achieved in multiple different ways will require the development of tools that can combine task analysis with demand assessments. It would also be useful to develop methods for assessing exclusion according to proportions within various different populations, such as different countries, specific market segments, or the workforce of a particular company.

1.6 References

Aldersey-Williams H, Bound J, Coleman R (1999) The methods lab: user research for design. Design for Ageing Network. Available at: www.education.edean.org (Accessed on 11 September 2007)

Clarkson PJ, Coleman R, Hosking I, Waller S (2007) Inclusive design toolkit. Engineering Design Centre, University of Cambridge, Cambridge, UK. Available at: www.inclusivedesigntoolkit.com (Accessed on 11 September 2007)

Dong H, Keates S, Clarkson PJ (2003) UK and US industrial perspectives on inclusive design. In: Proceedings of Include 2003, Helen Hamlyn Research Centre, London, UK

Grundy E, Ahlburg D, Ali M, Breeze E, Sloggett A (1999) Research report 94: Disability in Great Britain. Corporate Document Services, London, UK

Gulliksen H, Messick S (1960) Psychological scaling: theory and applications. John Wiley and Sons, New York, NY, US

Poulson D, Ashby M, Richardson S (1996) USERfit: a practical handbook on user-centred design for rehabilitation and assistive technology. HUSAT Research Institute for the European Commission

Keates S, Clarkson PJ (2003) Countering design exclusion. Springer, London, UK

Langdon P, Keates S, Clarkson PJ (2003) Developing cognitive capability scales for inclusive product design. In: Proceedings of the 14th International Conference on Engineering Design (ICED'03), Stockholm, Sweden

Martin J, Elliot D (1992) Creating an overall measure of severity of disability for the office of population census and surveys disability survey. Journal of Royal Statistical Society Series A, 155(1): 121–140

Martin J, Meltzer H, Elliot D (1988) The prevalence of disability among adults. Her Majesty's Stationery Office, London, UK

Microsoft (2003) The wide range of abilities and its impact on computer technology: a research study commissioned by Microsoft Corporation and conducted by Forrester Research Inc. Available at: www.microsoft.com (Accessed on 11 September 2007)

Steinfeld A, Steinfeld E (2001) Universal design in automobile design. In: Preiser W, Ostroff E (eds.) Universal design handbook. McGraw-Hill, New York, NY, US

WHO (1980) International classification of impairments, disabilities and handicaps. World Health Organization, Geneva, Switzerland

Chapter 2

Using Constraints in the Understanding of the Interactions Between Products and Humans

A.J. Medland, B.J. Hicks and S. Gooch

2.1 Introduction

The acceptability of consumer products often relies upon ease of successful use. Here there is a crossover between the functionality and how the user can derive the actions required. The form of these relationships can encompass the complexity of the physical actions that are required, the experience and capability of the user, and the social context in which the use takes place.

For example, in early work on vending machines it became obvious that differing approaches occurred or emerged between novice and experienced users. The new user was cautious, sometimes in the extreme. In work on the development of automatic petrol stations new users would read all instructions carefully and walk around the site before venturing in to part with their money and putting petrol in their tank. They were often looking for clues or familiar patterns that helped them recognise the tasks that would need to be performed.

The more experienced user would boldly approach the vending equipment making assumptions on how it operated, based upon previous experience (which could be helpful if it was the same type of station or very unhelpful if it was an alternate design).

The successful design of such products that rely upon close human interaction, where the operator is self taught, requires an understanding of both the approach of the human and of the product or machine operation. Clearly when the sequence and timing of the machine has been laid down it is not possible for the operator to change them. Instructions and feedback must thus be provided to allow the operator to understand the desired sequence of operations and most importantly what comes next. However, the derivation and inclusivity (with respect to user experience, capability and physical characteristics) of such instructions is, in general, fundamentally limited for all but the simplest of interactions.

One means to overcome such barriers is to explore user-product interactions during the early stages of the design of a product. However, achieving this is all but prevented by a lack of computational means capable of predicting the changes in user interaction that arise as a consequence of changes in the design of the product.

2.2 Modelling User-product Interaction

A considerable number of computing tools have been created to help different aspects of designing. In the area of product design these extend from styling packages through the classic geometric modellers to those handling manufacturing issues. To these can be added analysis tools such as ergonomics that allow certain aspects of the human interface to be evaluated. Two common ergonomic analysis tools are Sammie CAD (Sammie CAD Ltd., 2007) and Jack (Siemens AG, 2007).

However, these computer techniques can, in general, only be used sequentially, moving down from the concept through the detailing and on into the manufacturing. If, however, problems of an ergonomic nature are found at a late stage, much of the previous work may have to be revisited and changed.

Whilst considerable effort can be put into creating both computer-based and practical ergonomic studies neither provides the ability to assess and automatically correct errors during the early stages of design. Some areas of possible danger may only be identified accidentally or even after completion. An approach that integrates these different aspects of design creation and evaluation is thus required. The contribution of this paper is to report the creation of a constraint modelling approach for understanding human-product interaction.

2.3 Constraint Modelling

The constraint modelling approach seeks solutions to problems not by creating explicit equations but by undertaking searches across the problem space for conditions where all the constraints (rules) are true. For example, in the context of mechanical systems design, if two parts are required to assemble, then the rules of the fit are defined (such as a shaft must fit in a hole). This then defines the relationship between the diameter of the shaft and the hole. Either diameter (or even both) can be changed to obtain the required fit.

Within the constraint modelling approach the rules are first created (*i.e.* that which needs to be true) and the 'free' variables, for the search, selected. Direct search techniques are then applied to seek a true solution. Within such an approach there are no guarantees that a true solution actually exists or that a number of rules, although true, are not in conflict. The approach thus attempts to minimise the error in the total truth of the problem to an acceptable level.

The combination of rules and errors may indicate that no true or acceptable state can be generated. It is here that different strategies need to be applied. Some of the rules may not be solvable with the declared variables. Either the rules need to be redefined or new variables selected (or both). Such conflicts or trade-offs are common as the understanding of a problem increases and the design process evolves. In fact, the ability to creatively explore the problem space is central to generating understanding, particularly, where new or ill-understood problems are considered.

The fundamental need for designers to be able to redefine, revise, evolve and explore the constraint set (problem space) as understanding increases is one of the key benefits of using a constraint approach. The ability of constraint-based approaches to support the representation and exploration of the interactions between machines and materials, processing systems and products, and machines and operators has been demonstrated (Molenbroek and Medland, 2000; Hicks *et al.*, 2006a, 2006b).

2.4 Human Modelling Using Constraints

This constraint resolution approach has been developed over the last thirty years and been used extensively in the design and optimisation of manufacturing machines as well as other devices. In the last decade a human model was created based up on the anthropometric design assessment program system (ADAPS) model generated at the Technical University of Delft (Ruiter, 1999).

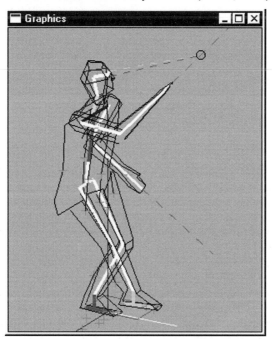

Figure 2.1. The underlying structure of the constraint-based manikin

Figure 2.2 shows a manikin attempting to work (touch) a specified point on a machine whilst also looking at that point, keeping both feet on the ground and maintaining balance (balance is defined by a rule that calculates the centre of gravity which the search routine attempts to keep within the base of support of the feet).

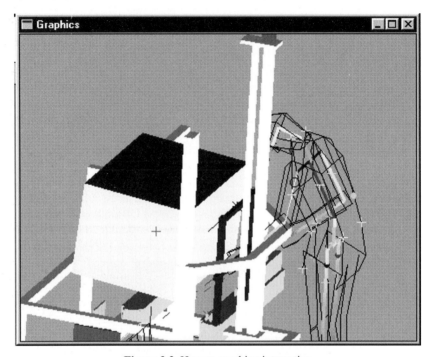

Figure 2.2. Human-machine interaction

Whilst the model of the manikin has more than eighty degrees of freedom, in practice, only a subset of these variables will control a particular posture or influence an interaction. In such studies the 'free' variables (joint rotations) of the manikin are automatically selected through the use of sensitivity analysis to identify the dominant variables and eliminate the insignificant ones. This process not only reduces the number of degrees of freedom, which improves computational efficiency, but also prevents the search from exploring redundant regions of the space. The latter of these is important to prevent the underlying numerical methods from unnecessarily altering variables which cannot subsequently be corrected as their contribution is negligible and hence cannot be measured. In this manner it is possible to predict human postures and movement patterns more accurately.

Should the manikin have difficulties in finding a satisfactory solution (even when all postures and positions around the machine have been explored) then new positions for the action point on the machine can be included. If for instance the point represents the position of a selected control, design work can be undertaken to find an alternative and better position for it. Thus a configuration satisfying both the human operator and the machine operation can be determined.

2.5 Predicting Postures Using Constraints

Many human studies commence with the model standing or sitting in a balanced state. These are often taken as a given condition but must be provided within any modelling environment. Within the constraint environment the position of the centre of gravity of the body at any instance can be calculated from masses associated with each part of the body. For the manikin to balance the centre of gravity must lie within the base of support, which can be calculated for standing as within the footprint or sitting as the seat-print.

In other more complex studies the centre of gravity will change as movements and new postures are demanded. One such activity is that of moving from sit to stand. Here the rules defining sitting and those for standing are different but must merge at an intermediate point where the force balance moves from being provided by the seat to being provided by the feet upon the ground. At this intermediate state both sets of rules are true unless momentum is used to 'throw' the body from one state to the other. The predicted movement pattern and intermediate posture for the sit-to-stand sequence is shown in Figure 2.3. The accuracy of the predicted movement pattern has been validated through practical studies and is reported elsewhere (Mitchell , 2007).

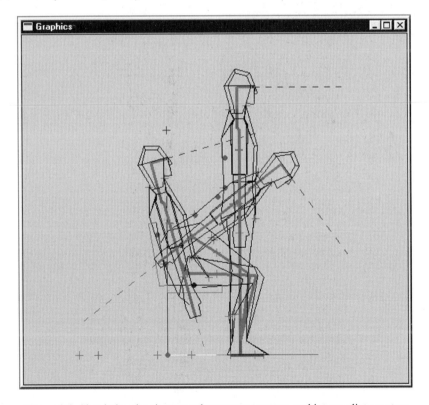

Figure 2.3. Simulating the sit-to-stand movement pattern and intermediate posture

2.6 Modelling the User-wheelchair Interaction for Improved Design

A study of the sitting posture of users in a wheelchair shows the close integration necessary between the individual and the designed product. The seating position must be comfortable allowing the person to sit in a relaxed 'driving' position with the feet on the support platform and the body leaning against the back rest. The user must also be capable of reaching the wheel push-rim over a range to be able to propel the wheelchair.

These rules have been incorporated into the modelling environment and the manikin set to find a position in which all actions and postures are successfully carried out. Studies currently being undertaken by colleagues in Christchurch, New Zealand, have shown that such postures are similar to those found from real experimental studies of both able-bodied and disabled people. See Figure 2.4.

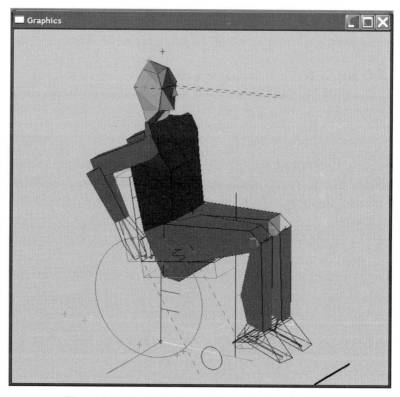

Figure 2.4. Simulating the seated posture in a wheelchair

The forces and power required under different conditions have been studied experimentally using a special purpose dynamometer in order that the effectiveness of the wheelchair design can be studied. It is intended that the limiting conditions will be compared directly with those provided by the constraint-based manikin.

Figure 2.5. Study of wheelchair propulsion being undertaken in New Zealand (Gooch)

As the models in the constraint environment are all parametric, certain design parameters, such as wheel size, balance point and seating height can all be declared as free variables in an optimisation search. The best configuration to meet the chosen conditions can thus be sought to meet these limiting conditions. By changing the geometric parameters of the manikin an optimised wheelchair can be investigated to meet either the needs of an individual or those of a specified user group.

2.7 Modelling to Improve Designs

The application of a constraint-based approach to design problems, such as the wheelchair study, allows the various interactions between humans and products to be studied in an integrated manner. The full human model exists within the constraint environment and rules can be generated to define the relationships with any suitably modelled product. A wide range of product interactions can be represented, including all aspects of touching, looking and operating together with those of human posture and balance. The ability to represent a range of interactions between user and product, and predict the changes in human posture and movement patterns brought about by changes in the product design, provide a unique modelling environment. In particular, user-product interaction can be considered during the early stages of design to create more inclusive products or during the redesign of existing products to optimise their function for a particular user or group of users both of which are not readily available through existing ergonomics modelling tools.

2.8 Conclusions

The requirement to be able to model and reason about product-user interaction during the early stages of design has been discussed. It has also been argued that one of the major limitations of commercial tools, in the analysis of ergonomics, is their inability to predict the changes in interaction brought about by altering the product design. Such limitations are a particular issue where interactions are dependent upon locomotion. In these cases, the ability to reliably predict human postures and movement patterns arising from a change in the designed environment or a change in user capability, such as ageing or disability, poses a significant research challenge.

To begin to address this research challenge, a methodology for modelling human-product interaction based upon constraint-based techniques has been presented. The overall approach and underlying numerical techniques are summarised and a number of modelling scenarios discussed. The scenarios include the complex motion of sitting-to-standing and the modelling and reasoning about the interactions between the user and the wheelchair for its improved design. In particular, in the case of the wheelchair it is shown how the design can be optimised to meet the needs of either individuals or selected groups of the disabled.

Following the success of the reported studies the approach is being extended to consider the requirements for modelling the operation of consumer produts to operate consumer products. These include the rules necessary to carry out actions such as pushing (for vacuum cleaner *etc.*), carrying objects (say up stairs) and working with the product in such scenarios as at a desk or in the kitchen.

2.9 Acknowledgments

The work reported in this paper has been undertaken as part of the EPSRC Innovative Design and Manufacturing Research Centre at the University of Bath (grant reference GR/R67507/0). The work has also been supported by a number of industrial companies and engineers. The authors gratefully acknowledge this support and express their thanks for the advice and support of all concerned.

2.10 References

Hicks BJ, Mullineux G, Medland AJ (2006a) The representation and handling of constraints for the design, analysis and optimisation of high speed machinery. Artificial Intelligence for Engineering Design, Analysis and Manufacture, (20)4: 313–328

Hicks BJ, Sirkett DM, Medland AJ, Mullineux G (2006b) Machine-material interaction and its impact upon the design of improved tooling. In: Proceedings of the 6th International Symposium on Tools and Methods of Competitive Engineering (TMCE 2006), Ljubljana, Slovenia

Mitchell RH, Medland AJ, Salo AIT (2007) Design methodology to create constraint-based human movement patterns for ergonomic analysis. Journal of Engineering Design, 18(4): 293–310

Molenbroek JFM, Medland AJ (2000) The application of constraint processes for the manipulation of human models to address ergonomic design problems. In: Proceedings of the 3rd International Symposium on Tools and Methods of Competitive Engineering (TMCE 2000), Delft, The Netherlands

Ruiter IA (1999) Anthropometric man-models, handle with care. In: Proceedings of the International Conference on Computer-Aided Ergonomics and Safety (CAES'99), Barcelona, Spain

Sammie CAD Ltd. (2007) The SAMMIE system a computer based human modelling tool. Available at: http://sammiecad.com (Accessed in October 2007)

Siemens AG (2007) Technomatix Jack – ergonomics and human factors product. Available at: www.ugs.com/en_us/products/tecnomatix/human_performance/jack/index.shtml (Accessed in October 2007)

Chapter 3

User Involvement and User Data: A Framework to Help Designers to Select Appropriate Methods

J. Goodman-Deane, P.M. Langdon, S. Clarke and P.J. Clarkson

3.1 Introduction

Many methods have been developed and adapted to help designers to understand, empathise with, and quantify users' situations, through both direct user involvement and more indirect use of user data. These methods vary widely, with different goals and suited for use in different situations. However, designers often find it difficult to select the most appropriate for their needs, often leading to inappropriate method use. We therefore propose a framework to help designers to make informed decisions about methods. The framework identifies the key information needs of designers in making these decisions, based on observations, interviews, card-sorting studies and a literature review. We further discuss how the framework may be populated, giving an example and discussing key issues.

In order to design products that will really be useful and usable in practice, it is important that designers keep in mind the users' needs and desires (Griffin, 1996). This is particularly important in inclusive design because of the special needs of many of the users and the extra challenges they face in using products. However, it can be very hard to do. Designers are often young and able-bodied and can find it difficult to understand and remember the characteristics of people in very different life situations with different needs, abilities and desires (Eisma *et al.*, 2003).

Many methods have therefore been developed both for involving users directly and for helping designers to understand and empathise with users' situations. These user-focused methods or "user methods" can be invaluable in enabling inclusive design. However, not all of them are suitable for all situations. For example, a designer may not have the resources to conduct a full-scale usability test nor be at a stage of design where such a test would yield valuable information.

Faced with the large number of user methods available and uncertainty over what they provide, designers often find it difficult to select the most appropriate ones for their needs. As a result, they may use unsuitable methods or, feeling that none of the methods are suitable, avoid user involvement or even user methods entirely. For example, Crilly and Clarkson (2006) explain that:

> *"Designers may feel that they themselves are representative of the target market, that they already have sufficient understanding of consumers, or that, in any case, consumer research is of limited value in yielding useful insights."*

We therefore present a framework to help designers select user methods. This framework will include a structured set of methods, and ways to access this set. In this paper, we focus on the structure of the set. We identify designers' key information needs in determining which methods to use, and show how the method set can be structured around these issues to aid selection. This set could be accessed in various ways, *e.g.* as a set of index cards or a searchable database. We plan to investigate this further in future phases of this research.

3.1.1 Related Work

Comparison with the Standard Approach

Our approach contrasts with that often taken in human factors, design and social science textbooks (*e.g.* Bryman, 2004; Stanton *et al.*, 2005), which focus on equipping the designer or researcher to carry out any of a set of methods, rather than on selecting between them. These books describe each method in detail, but there may be limited structure to the set as a whole, which is commonly ordered with respect to a key characteristic. Some further information to aid method selection may be provided, but, as this is not the main point of the database, it may be incomplete or unorganised, making it difficult to compare methods.

In contrast, we do not provide detailed descriptions of methods but instead focus on enabling comparisons between them: we concentrate on providing in a consistent form the important information for choosing between user methods. The two approaches are complementary. Our framework is intended to aid in deciding which user methods to use and why to use them for inclusive design. The standard texts then can be used to find out more about how to conduct them in practice.

Categorising Design Methods

While many method sets provide limited structure, several ways of categorising design methods have been proposed, many with the purpose of helping designers to select appropriate methods. However, these categorisations, while providing helpful guidance, are often based on theory rather than an understanding of the specific needs of designers. For example, van Kleef *et al.* (2005) propose a categorisation based on consumer psychology and marketing and López-Mesa and Thompson (2003) describe methods on a spectrum based on "principles from the field of creativity that concern the way people think". Others do not fully explain why they chose the particular dimensions used in their categorisations (*e.g.* Hom, 1998; Design Council, 2007).

We feel that it is important to provide a framework that is based more firmly in design practice, identifying the issues that are important for method selection in practice. Aldersey-Williams *et al.* (1999) went some way towards this in the development of MethodsLab, a directory of user methods. They sought to position methods along "two axes that reflect designers' concerns" and developed the directory through regular meetings with designers and researchers. The result is a very valuable resource, reflecting some real concerns, and listing several important issues that are also included in the framework in this paper. However, they did not examine the wider picture of design practice.

This work was later built into a more holistic resource for inclusive design (the RSA Inclusive Design Resource, now (EDeAN, 2007)), based on investigations of designers' needs for inclusive design. This resource included descriptions of many user methods, but did not attempt to categorise them.

Our work builds on and enhances this previous work by basing a framework for method selection on wider and more systematic studies of design practice and designers' preferences.

3.2 Methodology

The framework was based on findings from several studies that examined different aspects of the use and categorisation of design methods, as described below. Each study identified various issues of importance in choosing methods. We selected the issues that were common across several studies, as well as ones that individual studies highlighted as being especially important, reflecting the particular insights from those studies. These were then combined into a single set of method characteristics that are important when selecting methods, and formed the basis of the framework described in the following section (Section 3.3).

3.2.1 Triangulated Study of Design Practice

We conducted a triangulated study, using multiple research methods, to obtain a rounded picture of design practice. The study incorporated a literature review, an observation and interview study of six teams of designers involved in an inclusive design competition, interviews with two design experts, and a survey of response to inclusive design in industry. The study examined various aspects of design practice and inclusive design, including the use of design methods and of information. More details of the study and some initial results can be found in (Goodman *et al.*, 2006a).

Amongst other findings, we built up a picture of how design methods are used in design practice, particularly focusing on methods of user involvement and the use of user information. We identified methods that are more and less likely to be used, and determined what their characteristics are. Designers were also asked directly what would encourage or discourage them from using a method. From these, we identified issues of importance in determining whether a method is used. The kinds of methods used in different parts of the design process and for different purposes were also identified.

3.2.2 Card-sorting Study of Design Methods

A card-sorting study was conducted with 21 product and communications designers. Each participant was given a set of 57 cards, describing a range of design methods and techniques. The methods were selected from a larger set of over 330 design methods, identified through a literature survey. Both individual papers and sets of methods were examined (*e.g.* Aldersey-Williams *et al.*, 1999; IDEO, 2003; Stanton *et al.*, 2005). The methods were chosen to cover a representative range, with an emphasis on methods for involving and understanding users, since this is of particular importance in inclusive design.

To avoid influencing participants' decisions, we asked them to group the cards, based on their design knowledge and experience, using any criteria they liked to do so. An example of the result is shown in Figure 3.1. They then rated the methods according to their frequency of use and how much they thought they would enjoy using them. The process and pilot results are described in (Goodman *et al.*, 2007).

The card-sort indicated designers' underlying perceptions and understanding of design methods and the considerations they feel are important in categorising them. The method ratings showed how these issues impact method use in practice.

Figure 3.1. A completed card sort from the study

3.2.3 Literature Review

A review of the literature examined other categorisations of design methods, both of methods in general and of user methods in particular. We compiled a list of issues that were used as the bases of these categorisations. From this, we extracted the most commonly identified issues. We also noted issues specifically identified by those studies that were based on an understanding of the design process, as opposed to studies suggesting issues based on other theory or without justification.

3.3 The Framework

These studies identified a set of method characteristics of importance in selecting methods. This set is the basis of the framework, shown in Table 3.1. The table explains how each characteristic can be described and gives a possible (abbreviated) entry for the user method "informal interview". The following subsections describe the characteristics in greater detail.

3.3.1 Input Needed

Our studies identified cost (time and money) as a key reason why many methods are not used in practice. In fact, for many designers cost seems to be the main aspect influencing method use. Additional reasons included difficulties with the logistics of user involvement, especially with obtaining users, and fears, on the designers' part, that they lack the expertise to conduct user research properly. It is therefore important that the framework includes information on a method's cost and on its requirements for access to users and to specialist knowledge and skills. Requirements for user access should include information on the type of users and user contact (*e.g.,* face-to-face or more distant contact, with mainstream, extreme or boundary users), as these also affect the cost of user involvement.

However, many methods can be used in multiple ways and thus do not have a single fixed cost or set of requirements. The framework therefore needs to describe alternative ways of using a method and their impact on other issues such as cost.

3.3.2 What is the Method Suitable For?

Design Stages
Textbooks on design methods often divide up method sets according to the design stage at which each method is most applicable (*e.g.,* Baxter, 1995). We too found that different methods tend to be used at different stages of the design process, indicating that information on this would help in selecting appropriate methods.

Our study of design practice found that different companies use different sets of stages, and that designers use stages flexibly and iteratively. Nevertheless, it was possible to construct a rough outline of top-level stages. These do not apply to all situations, but could help many companies in considering which methods are suitable for them at different points in design. The stages were:

- briefing, defining the problem or opportunity;
- analysis of the problem, information search and research;
- creativity, idea generation and synthesis;
- design development, including prototyping and some evaluation;
- finalisation of the design;
- manufacturing and delivery.

Table 3.1. The proposed framework, with an abbreviated example

Issue	Details	Example: Informal Interview
Input needed		
Cost	- Overall cost estimate - Optionally, time, money & personnel	2.6 (on a scale from 1 to 7, 1: cheap, 7: expensive)
Access to users	- What type of contact is needed? - Length and number of sessions - How many users? - What kind of users? (Mainstream, extreme, boundary)	Face-to-face or distance contact 1 or more session per user, each typically 10-60 mins Varying, with differing accuracy and cost. 5-10 users is common. Different kinds of users for different purposes.
Specialist resources	- Knowledge, skills and training - Equipment - Other	Good social and listening skills. Recording equipment is not necessary but can be useful.
Flexibility	Alternative ways in which the method can be used	Variable formality and structure. More informal are simpler to organise but more formal help to ensure useful data.
What is it suitable for?		
Design stages	Which design stages are suitable, chosen from the list in Section 3.3.2. - For each, its degree of suitability (Low, Medium, High)	- Briefing: High – uncover opportunities - Analysis: High – learn about problem - Creativity: Low – suggest ideas to address problem - Design develop.: Med – give feedback
Usage goals	Key design goals that the method helps to achieve, chosen from the list of possibilities in Section 3.3.2. - For each, its degree of suitability	- Producing ideas: Low - Evaluating ideas and prototypes: Med - Understanding user needs in a particular area: High
Output	Type of output Quality of output (validity, reliability and generalisability) How easy is it to act on the output?	Qualitative, open-ended info re. users. May be some interviewer/subject bias. Generalisations are not reliable. Depends on the questions asked but usually requires interpretation (which may be time-consuming).
Type of method		
Type of method	Type, chosen from a list of possibilities (see Section 3.3.3)	- Understanding users (direct contact) • Interview-based techniques
Practical information		
Related methods	Common variants and other names for the method	"Light-weight", unstructured or semi-structured interview
Description of the method	Description Link to further information	An informal, open discussion with the user. It does not necessarily have a set list of questions. (Bryman, 2004)
Summary of strengths and weaknesses	The particular strengths of the method Its weaknesses	Simple and easy to conduct, provides detailed information from real users. Not generalisable and may be biased.

A method's entry for the issue "Design stages" thus contains a list of the stages (selected from those above) where the method is most useful, together with an indication of its degree of suitability for each of these stages.

Usage Goals

Different methods are suitable for achieving different goals. For example, CAD modelling is great for visualising and developing ideas but not very well suited to understanding the market. If designers are to select the most appropriate methods, they therefore need to know which goals each method helps to achieve.

To make this information more consistent and easier to navigate, we draw the goals from a common list (while also providing an "Other" category, for cases whose key goals do not match the given options). Our studies identified the following goals as being of particular interest to designers. Although some of these are not directly related to users, user methods can still help to achieve them, and so they are included in the list of possibilities. For example, talking to users can be a rich source of ideas, as the users describe creative ways in which they have addressed difficulties in everyday life:

- understanding the market (the competition, business and population as a whole, as opposed to individual users);
- producing ideas;
- analysing and selecting ideas;
- visualising and developing ideas;
- evaluating ideas and prototypes.

In addition, we identified the following key objectives of user methods, based on our expertise and previous studies in inclusive design (*e.g.* Clarkson *et al.*, 2007). By including these as well as the above in the framework, we address particular needs in user-centred and inclusive design that may not be obvious from studies of general design practice. Including these also helps to make designers aware of the potential of user methods:

- giving an overview of population diversity and inclusive design issues;
- providing an understanding of user needs in a particular product area;
- providing detailed data on users' abilities, to support detailed design;
- predicting market performance (including an estimate of how many and who will be excluded from product use).

Output

Methods produce different kinds of output. For example, results may be quantitative or qualitative; and allow flexible interpretation or have defined, concrete meanings. Our studies found that designers generally prefer methods that produce open-ended, inspirational data, but they also value more detailed, quantitative data for some parts of design. The type of output is therefore an important concern in selecting methods, and should be included in the framework.

Designers also need to know how the results can be used. They need information on how reliable and generalisable the results will be, so that they do not try to apply them in inappropriate ways, nor become frustrated when they later find they cannot generalise the information they went to so much trouble to obtain.

In addition, some types of output are easier to act on than others. For example, Boztepe (2007) explains how designers often find it difficult to translate "the very broad, general, and rich information about people" obtained from ethnographic research into "an *actionable* form". Designers would therefore benefit from information in the framework on how "actionable" the results are and how they can be turned into implications for design.

3.3.3 Type of Method

The results of the card-sorting study (Section 3.2.2) reveal groupings that designers themselves use in thinking about design methods. Including these in the framework helps it to fit more closely with designers' views of methods and thus makes it easier for them to find appropriate methods for their needs.

The study found the following key categories, which are connected to a method's goals but are also related to the techniques used and the ways in which users are involved. Thus they most closely reflect a method's "type":

- analysis (analytical methods for examining situations, ideas and products);
- concept design (idea generation and selection);
    - idea generation;
    - sketching and introspection;
- examining the market (market research methods);
- prototyping;
- understanding users without direct user contact;
    - describing users;
    - remote contact with users;
    - user sampling strategies;
    - role play;
- understanding users through direct user contact;
    - interview-based techniques;
    - product evaluation;
    - other methods for user contact.

3.3.4 Practical Information

In addition to the issues listed above, designers need information on the method itself and how to run it. We therefore include a description of the method and a link to further information. A list of alternative names for the method and its common variants is also provided, since methods are often known under other names and in other forms. Lastly, we include a summary of the method's strengths and weaknesses. Although many of these can be determined by examining other issues in the framework, this further brief summary helps in quick method selection.

3.4 Populating the Framework

We plan to populate this framework with a selection of methods for understanding and involving users, covering a range of different method types and including both commonly used methods and ones that are particularly useful for inclusive design. Relevant methods from the card-sorting study will be used as the basis for this set.

The entry for a method will contain information on each of the issues in the framework, as shown in the abbreviated example in Table 3.1. Some of this information is provided by the studies described in Section 3.2, while other data comes from the literature, our background knowledge and continuing work.

For example, the overall cost estimate given in Table 3.1 was obtained as part of our card-sorting study, where five design experts rated the cost of the 57 selected methods, using a seven-point Likert scale. The method type was also obtained through the card-sorting study: cluster analysis was used to identify which category each method was commonly placed in (Goodman *et al.*, 2007).

Information from the card-sorting study can also be used as the basis for the method descriptions. As part of this study, one-paragraph descriptions of each method were written and cross-checked by other experts in the field. These can be expanded to form more comprehensive versions for the final database.

3.5 Overarching Issues

3.5.1 Combinations of Methods

This framework provides information about each method separately, helping designers to choose individual methods. However, in practice, designers use several methods on a project. We therefore plan to enhance the framework with guidance to help designers select sets of methods that work well together, as well as ones that are individually suitable. It is important to get a good balance of different kinds of methods to provide a balanced view of the user and the design problem, and to support different parts of design. Of particular importance for inclusive design is the balance between using second-hand user data and direct user involvement. These have different roles and achieve different purposes. User data is often cheaper to use and can provide a larger scale picture of the user group. However, direct user involvement often creates greater empathy with the users and can be more inspiring (Coleman *et al.*, 2003). We plan to include within the framework a discussion of these different roles of user data and user involvement.

3.5.2 Flexible Method Use

As mentioned in Section 3.3.1, methods can often be used in more than one way, *e.g.* with varying degrees of formality or different kinds of interview questions. These differences result in variations in the cost of the methods and in the type and reliability of the results. Our studies have found that such flexibility is important to designers, who often adapt methods for their own use (Goodman *et al.*, 2006b). It is therefore important that this flexibility be captured in the framework.

One way to do this is to list each variation as a separate method in the framework. However, this does not adequately convey the flexibility of the methods and does not capture all of the possibilities, since variations often lie on a spectrum, without clear distinctions between them. In addition, this would lead to a combinatorial proliferation of methods, resulting in a large and unwieldy database.

Another possibility, adopted in our framework, would be to allow a range of values for each entry. The database can then be searched for methods whose ranges fall within the designer's requirements. However, care would need to be taken not to describe too many variations in a single entry because this could lead to confusion and make it difficult to identify the impact of changing one aspect of a method (*e.g.* level of formality) on another (*e.g.* cost).

Therefore, a good level of granularity is needed, with a manageable number of distinct methods, each covering a limited number of variations. In addition, we are investigating further ways of expressing the impact of changing a method, perhaps using interactive database features. For example, selecting a particular value for one aspect could set other aspects to their related values.

3.6 Conclusions and Further Work

Designers often find it difficult to select the most appropriate user methods. We have therefore proposed a framework to help them make these decisions, identifying their key information needs, based on studies of design practice.

We are currently populating this framework, including methods both for involving users directly and for helping designers to understand and empathise with users' situations more indirectly. As the framework is populated, it will be adapted and improved to best express this range of methods.

We will consider different ways in which designers can access the method set, such as through a set of index cards or a searchable database. In developing the interface, we will take into account designers' preferences for information formats, striving to present the information in a visual, attractive and simple way. We plan to consider how flexibility in method use can be best expressed, perhaps by using interactive database features. We also plan to enhance the framework with guidance to help designers to select sets of methods that work well together, as well as ones that are suitable individually.

Populating the framework will also help to identify shortcomings in the methods currently available. For example, our studies identified time and budget as the key restrictions for method use. However, designers often feel that there is a lack of cheap yet reliable user methods. We hope to identify and propose ways of meeting these inadequacies.

We plan to develop the database through ongoing consultation with and feedback from designers, determining whether it does make method selection easier and how it can be improved. As a result, we hope to provide a complete and well-structured set of user methods that will enable designers to easily and effectively identify appropriate ones for their needs.

3.7 References

Aldersey-Williams H, Bound J, Coleman R (1999) The methods lab: user research for design. Available at: www.education.edean.org/pdf/Tool039.pdf (Accessed on 7 September 2007

Baxter M (1995) Product design. Nelson Thomas, Cheltenham, UK

Boztepe S (2007) Toward a framework of product development for global markets: a user-value-based approach. Design Studies, 28(5): 513–533

Bryman A (2004) Social research methods, 2nd edn. Oxford University Press, Oxford, UK

Clarkson PJ, Coleman R, Hosking I, Waller S (eds.) (2007) Inclusive design toolkit. Engineering Design Centre, University of Cambridge, Cambridge, UK. Available at: www.inclusivedesigntoolkit.com (Accessed on 7 September 2007)

Coleman R, Lebbon C, Myerson J (2003) Design and empathy. In: Clarkson J, Coleman R, Keates S, Lebbon C (eds.) Inclusive design: design for the whole population. Springer, London, UK

Crilly N, Clarkson PJ (2006) The influence of consumer research on product aesthetics. In: Proceedings of the 9th International Design Conference (DESIGN 2006), Dubrovnik, Croatia

Design Council (2007) Design methods. Available at: www.designcouncil.org.uk/en/About-Design/Design-Methods (Accessed on 7 September 2007)

EDeAN (2007) Design for all education and training website. Available at: www.education.edean.org (Accessed on 10 September 2007)

Eisma R, Dickinson A, Goodman J, Mival O, Syme A, Tiwari L (2003) Mutual inspiration in the development of new technology for older people. In: Proceedings of Include 2003, Helen Hamlyn Research Centre, London, UK

Goodman J, Langdon PM, Clarkson PJ (2006a) Providing strategic user information for designers: methods and initial findings. In: Clarkson J, Langdon P, Robinson P (eds.) Designing accessible technology. Springer, London, UK

Goodman J, Langdon PM, Clarkson PJ (2006b) Equipping designers for inclusive design. Gerontechnology, 4(4): 229–233

Goodman J, Clarke S, Langdon P, Clarkson PJ (2007) Designers' perceptions of methods of involving and understanding users. In: Proceedings of the 4th International Conference on Universal Access in Human-Computer Interaction (UAHCI 2007), Beijing, China

Griffin A (1996) Obtaining customer needs for product development. In: Rosenau et al. (eds.) The PDMA handbook of new product development. John Wiley and Sons, New York, NY, US

Hom J (1998) The usability methods toolbox. Available at: http://jthom.best.vwh.net/usability (Accessed on 3 September 2007)

IDEO (2003) IDEO method cards. IDEO, Palo Alto, CA, US

López-Mesa B, Thompson G (2003) Exploring the need for an interactive software tool for the appropriate selection of design methods. In: Proceedings of the 14th International Conference on Engineering Design (ICED'03), Stockholm, Sweden

Stanton N, Hedge A, Brookhuis K, Salas E, Hendrick H (2005) Handbook of human factors and ergonomics methods. CRC Press, Boca Raton, FL, US

van Kleef E, van Trijp HCM, Luning P (2005) Consumer research in the early stages of new product development: a critical review of methods and techniques. Food Quality and Preference, 16(3): 181–201

Chapter 4

Engaging the Ageing: Designing Artefacts to Provoke Dialogue

S.J. Bowen and P.M. Chamberlain

4.1 Introduction

This paper discusses ongoing work by the authors on how the design and deployment of artefacts can be used as a method of engaging stakeholders in research. A case study, *Living rooms*, is described where we used "critical artefacts" to develop our understanding of the design of the home to support independence and wellbeing in later life. Through the project we have developed key themes and design concepts that will directly inform further work on designing for tomorrow's older people. We describe several examples to demonstrate the effectiveness of our method.

Living rooms will also inform the design of a "test lab" for exploring design concepts and theories with stakeholders as part of the *Lab4Living* research collaboration. In particular, we describe the social and physical factors that influenced stakeholders' engagement.

In addition to demonstrating the value of one particular approach, we aim to contribute to a broader discussion of the role and engagement of "users" in practice-based design research.

4.2 Inclusivity

While the drive towards inclusive research is bringing researchers, practitioners, end users and industry together in new and exciting multi disciplinary ways, the degree and extent of involvement varies widely, with many diverse research approaches labelled as "inclusive". A widely accepted principle of both the research and practice of design is that it is concerned with building and adapting environments and products in a manner that makes them appealing and functional to all ages and abilities, thereby mainstreaming access for disabled or the elderly rather than making them exceptional. However there are conceptual problems and risks with inclusive design if a narrow interpretation is adopted. Inclusive design can be misunderstood to mean a "one size fits all approach", which is not in accord with contemporary consumer theory that suggests that people seek to define and express themselves as individuals through the products and services they buy. A seemingly contradictory position is expressed in the Markets of One concept of Gilmore and Pine (2000) which takes the concept of market segmentation to its logical conclusion, by describing how to meet the needs and requirements of individual consumers. While the principles of design for all and markets for one might appear as polar opposites, the emerging "mass customisation" technologies and user appropriation strategies (McCarthy and Wright, 2004) present the possibility of products and services which may better serve the individual needs and aspirations of consumers across a wide range of ages and abilities.

To help us understand these individual needs it is important to gain a good understanding of the "user": hence the growing trend to involve "users" through user-centred research methodologies. User-centred design is a widely accepted term popularised by Norman (1990) and others (McDonagh-Philip and Lebbon, 2000). However, there has been some recognition of the limitations of user-centred design in recent design discourse. User-centred design implies that it is only the *users* of a product who are considered in its design. But there are numerous others affected by a product such as those who manufacture, transport, install, configure, maintain, and ultimately dispose of it. "Human-centred" design is a development of the idea of user-centred design and attempts to include this more diverse set of "stakeholders" in the design process. The focus on people merely as users can lead to "usability" being the prime goal of design. Usable products have been designed so that their function is easy to achieve, but the underlying intention of this function is rarely engaged with in the design activity. Design can consider more than a product's utility, marketability or profitability. Buchanan (2001) has argued that design should be grounded in a more fundamental principle – to support or advance human dignity.

4.3 Research by Designing

The authors work in the Art and Design Research Centre (ADRC) at Sheffield Hallam University (SHU). A common theme of ADRC projects is the use of designing and making as methods of enquiry (Chamberlain *et al.*, 1999; Walters *et al.*, 2003; Walters *et al.*, 2004, Walters, 2006). Rust (2004) has shown how designers' creation of "new worlds" can open up new contexts for research. We are developing an approach where designers' creation of artefacts is key. We have previously discussed how such artefacts can facilitate communication between multidisciplinary research teams and engage stakeholders with novel situations; artefacts as vehicles for knowledge and prompts for reflection (Chamberlain and Bowen, 2006).

We develop our position here, suggesting that artefacts can be used to both further an enquiry and express the understanding an enquiry has developed. Central to this approach is that stakeholders and designers jointly engage with the artefacts.

This iterative engagement informs the designers' understanding of the context and the development of further artefacts. Crucially this engagement is part of a design activity rather than a research activity. The designers' understanding emphasises an holistic view of the challenges and opportunities of designing for a context, rather than explicit knowledge about that context. But if the engagement serves the activity of designing, where does the research happen?

We suggest that the designer develops insights about the context in participating in these activities. These insights are generally internal but are made implicit in any subsequent artefacts designed. The artefacts are the contribution to knowledge and can inform further research. Rust (2007) provides a wider discussion of these ideas.

4.4 Designing for an Ageing Population

It is well documented that the increasing proportion of older people in the UK will bring about significant future care issues (Ladyman, 2005). In response to this significant demographic change the authors were awarded funding through the Strategic Promotion of Ageing Research Capacity (SPARC, 2007) to investigate whether design can help maintain independence at home in later life, the project titled *Living rooms*.

Many current homes are subject to costly adaptations and "add ons" that can stigmatise older inhabitants, but we believe solutions should consider dignity, choice and independence. Central to our study was a concern not just about survival (bells, alarms and whistles for when people are in danger) but an investigation into things that delight us and enhance our quality of life through wellbeing. Our approach considers older users as informed, demanding consumers rather than simply as people with disabilities in need of treatment. The latter 'medical model' (Newell, 2003) approach can portray older people as tragic victims in need of constant care.

It is also important to consider that today's older people reflect particular generational viewpoints, such as a "put up and make do" attitude inherited from wartime experience. We aim to design for "tomorrow's older people" who are likely to have very different aspirations and lifestyles not currently catered for.

4.5 Case Study: Living Rooms

The primary aim of the *Living rooms* project was to inform the design of a "test lab" for exploring design concepts and theories with stakeholders regarding the design of the home for changing healthcare needs, lifestyles and aspirations, with an initial focus on older people. A key objective was therefore to understand the value of our artefact-centred method of engaging with stakeholders.

Secondarily the project served as a scoping study. Continuing our ongoing investigation into designing for an ageing population, we aimed to identify key themes and issues for further research and design concepts relevant to stakeholder needs for development into new products and systems.

4.5.1 Participants

To ensure a more complete appreciation of the context was developed, participants were chosen to reflect a broad range of stakeholders. Consequently groups of between six and eight participants were recruited from four broad stakeholder categories: "active old", "frail old", "future old" and carers.

We anticipated difficulties in defining and recruiting these groups before the study began. What defines someone as old? What defines frailty or "active-ness"? Our approach was to select participants whose circumstances meant they would be likely to have the types of experiences and needs we wanted to inform our design understanding. For example the "frail old" group were residents of an apartment block that provides extra care facilities. It was therefore reasonable to expect several of them to have more advanced health care needs than older people living independently. However in reflecting on the workshops, we are careful to recognise that the group names are purely "placeholders" and do not prescribe the characteristics of their members. For example, these are the views of residents of an extra care housing scheme, not of "frail" older people. The groups participated separately in the research.

4.5.2 Method

The core tactic of our engagement with stakeholders in this project was our design and deployment of artefacts. The artefacts were provocative design concepts in the developing tradition of critical design (Dunne and Raby, 2001; Z33, 2007). They were intended to provoke critical reflection on the context rather than being realistic design solutions. These "critical artefacts" were presented via a series of images that portrayed fictional situations of the artefacts in use.

The stakeholder engagement took the form of a series of three discussion workshops. These workshops were presented to participants as an ongoing "dialogue" between the participants, as stakeholders, and the researchers, as designers. In the first workshop participants were asked to share their experiences with the researchers. In the second and third workshops the researchers "replied" via two series of design concepts which prompted further discussion with the participants. Each workshop lasted for around one hour and was videotaped.

Our participation in and reflection upon the workshop discussions, as designers, served to develop our understanding of the context. We expressed this understanding via the artefacts presented which then became tools to further our enquiry – to provoke discussion in subsequent workshops.

(For a larger discussion of this method and its relationship to critical design and related approaches, see Bowen, 2007.)

4.5.3 Influencing Factors

In conducting the study, we noted several factors that could influence participants' engagement with the artefacts. We will note these briefly below.

Relationships Between Participants
The personal relationships between participants had an impact on the "open-ness" of the discussions that took place. This was also related to the way in which the groups were recruited. For example the "active old" group were part of an older persons' community group who meet monthly, and the majority of the participants also take part in a weekly fitness session which preceded all the workshops. They could reasonably all be said to be friends, and are used to meeting each other in a situation centred on enjoyment and socialising. The carers group were all members of Sheffield's 'Expert Elders' network – a group of older people, recruited, trained and managed by Sheffield council representatives aiming to 'make sure older people influence how care and other services are designed to keep them healthier and active for longer' (Sheffield Partnerships for Older People Projects, 2006). Members of this network are consulted by commercial companies and the local authority in the development of products and services. The participants are accustomed to offering opinions based on their experiences and having this consultation valued.

The "active old" group's personal relationships and expectation of meeting for enjoyment made them more open and imaginative in the discussions, whereas the carers group's perception of themselves as "experts" and expectation of business-like consultation made them less open in discussions.

Workshop Settings
The physical settings of the workshops also had an impact. The "future old" workshops took place in meeting rooms within the same institution that the participants worked for, although these rooms were not in the same building as their own workplaces. The rooms therefore had a definite "work" connotation to them. Contrastingly the "active old" workshops took place in the hall of a small church. This space was used for numerous activities connected with the church and local community with evidence of the same: musical instruments, children's paintings, stacking chairs and a dance floor area. The hall did not have "work" or "home" connotations, but nevertheless had an informal "sociable" feel.

Expectations
Participants involved in design research tend to have different types of expectation from those engaged in more traditional data gathering where the research outcomes are often neatly packaged as a quantitative body of knowledge and information. The participants all expressed an interest in 'what we were designing'. Our experience suggests participants accept pure information as an outcome of "traditional" research but are more likely to expect newly-designed "things" from design research. However, design research activity rarely has a direct link from stakeholder engagement to concrete new designs or a short timescale in the

development of these designs. Participants' expectations therefore need to be carefully managed lest they have broader or even no expectations of where the study is likely to lead, and the form of its output. Careful briefing, training, continuity to support iterative development are important but not extensively demonstrated in much research. This can provide greater understanding for the stakeholders in their role but does often require a significant level of commitment by the participants. This in turn raises issues about recognition and rewarding stakeholders for their participation.

4.5.4 Resulting Design Insights

The primary output of this project is proposals for the design of our "test lab", notably the use of artefacts to engage with stakeholders. However, what is perhaps more relevant to the discussion here is to evidence the effectiveness of this method of stakeholder engagement. The study resulted in the development of our (designers') understanding of the context. This understanding can be expressed via both explicit and implicit insights: key themes we have identified for further research; and design concepts that express these themes together with other, implicit insights. A few examples of key themes and the third workshop design concepts expressing them are described below. Again it is worth noting that these design concepts are not intended as realistic product proposals, rather they embody our thinking and highlight opportunities for further product development.

In presenting these insights, we acknowledge that they reflect our design understanding at a given point in time. This understanding is evolving as our work continues. Also we acknowledge that these insights were not produced "in isolation". Although primarily informed by our participation in the workshops, they are also influenced by our own interests, previous and ongoing design research outside the SPARC project, and other experiences in the "wider world".

Key Theme: Wellness Monitoring
The majority of systems for monitoring older people are about alerting when something is wrong: fall alarms, intruder alarms, bed-wetting alarms, wander alarms *etc*. Such systems only communicate when there is a problem. There is an opportunity to develop systems that monitor and communicate when there is not a problem, when everything is ok. These would be "wellness" monitors.

With alarm systems ("problem" monitors), the monitored person's wellness could be inferred by a lack of any alerts. However a lack of alerts could also be because of a failure in the alarm system. "Wellness" monitors would provide reassurance to both the person being monitored and those doing the monitoring via an explicit communication of wellness. They also provide a more continuous personal link between their users – they are reminded of each other more frequently ("I'm ok", "you're ok") rather than (hopefully) infrequently when something is wrong.

Key Theme: Minimal Interfaces

Complex interfaces for products and systems may cause problems. Firstly, the interfaces dictate what users do with the products – users tend to perform only the tasks the interface supports or affords. Secondly, users mistake the situation an interface represents for the real situation – mistaking that a fall detector warns of falls, when it actually warns that its gyroscopic switch has been activated, which may be due to bending or tipping not related to an actual fall. And thirdly, people absolve themselves of responsibility for verifying the progress and completion of tasks the products and systems support – "the fall detector is watching out for Granny, so I don't have to call round". The technology can become a "crutch" – it is relied upon unthinkingly.

Products and systems with minimal, almost primitive, interfaces may not readily cause such problems. An interface that provides very basic, abstract information and functionality does not prescribe the product's use; its users must define the product's function by appropriating it into their existing practices. A minimal interface is not easily mistaken for the real phenomena it is intended to represent – users must interpret the meaning through their experiences of using it in their own circumstances. And finally, because a minimal interface needs to be interpreted based on its users' appropriation of it in their own practices, the users cannot absolve themselves of responsibility for task verification. The users are fundamental to the use of the product or system. The product supports the user in completing the task; the product cannot complete the task without the user to "make sense" of what it is doing.

Design Concept: Glow Gems

The *Glow Gems* system consists of movement detectors communicating with small gem-like electronic devices. The movement detector is placed in a position in an older person's home that they regularly move through, for example a hallway or living room. When movement is detected, a signal is sent to the gem device which lights up (perhaps via a mobile phone using Bluetooth®). The light then changes colour slowly over time (from green to amber to red). The gem-device could be incorporated into numerous objects regularly worn – a cufflink, a pendant, a watch strap.

The system would give a friend or relation of an older person the means to be alerted of activity in the older person's home. Being something that is worn, the gem device can easily be checked throughout a busy day, irrespective of location. But the system cannot function alone without some intelligent and dynamic human intervention. For instance the gem-wearer must know if the older person is meant to be at home on particular days and times.

In devising this concept, we explored approaches to monitoring that were less intrusive of individual privacy. We also liked the deliberate "open-endedness" of the concept – that its users should appropriate and define its use. The "future old" group suggested using *glow gems* to 'play games' to check if the observer of the gems are 'really paying attention'.

This concept was inspired by the *virtual intimate object* developed by Joseph Kaye for communicating intimacy between couples in long distance relationships (detailed in Sengers *et al.*, 2005).

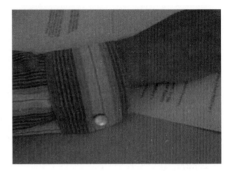

Figure 4.1. *Glow gems*

Key Themes: Secondary Conversations

Generally telephone conversations have a particular form. The communication is continuous and the conversation continues for as long as each party has something to say (and the time to listen). The communication is defined, deliberate and usually the primary focus of the participants' attention. But there are other types of conversation that are intermittent, undefined, unplanned and less attended to. These generally occur when the conversation is not the primary activity – it is secondary to another activity or on a par with it. For example at a coffee morning or around a family dinner table – the activities of sharing coffee or a meal provide the "purposes" of the social encounters, the conversation flows around them. There is an opportunity to develop product systems to support these sort of "secondary conversations".

Design Concept: Biscuit Tin Buddies

Biscuit Tin Buddies is a system to allow people to maintain social interaction with friends and family where distance or mobility make physical meetings difficult. The lids of special biscuit tins contain microphones, loudspeakers and wireless communication technology. When the lid of a tin is lifted, it opens an audio link with another "paired" biscuit tin in another location (perhaps using voice over IP). Meeting over "tea and a biscuit" is a familiar concept for British people. Therefore two groups of people can stay in touch by agreeing between them a regular time for tea and biscuits – the latter kept in their biscuit tins.

Figure 4.2. *Biscuit tin buddies*

4.6 Conclusions

We have discussed how our design and deployment of "critical artefacts" has proved a useful method of engaging with stakeholders and consequently developed our understanding of designing for tomorrow's older people. Bowen's PhD work is developing and documenting a methodology to support this method, for which *Living Rooms* has provided valuable evidence. And our increased understanding is also directly informing further research projects.

In a commitment to innovative and powerful inter-disciplinary collaborations that seek to develop new ways of involving users and building on their experience and track record, Professor Paul Chamberlain and Professor Gail Mountain (Director of the Centre for Health and Social Care Research at SHU) have recently set up *Lab4Living* (www.Lab4Living.org.uk) a research collaboration between Design and Health and Social Care at SHU. The initiative will operate from a brand new user lab in which experimental spaces are created to conduct both qualitative and quantitative studies of human behaviour and their interaction with the built environment. *Living rooms* has enabled us to identify issues posed by ageing, propose concepts and potential solutions, and reflect on the nature of how we engage users in the activity of designing.

The key themes and design concepts presented here are only a few samples of the outputs of *living rooms*. They also represent our understanding at a fixed point in time, an understanding that is evolving and developing through the practice of design in further projects.

4.7 References

Bowen SJ (2007) Crazy ideas or creative probes? Presenting critical artefacts to stakeholders to develop innovative product ideas. In: Proceedings of EAD07: Dancing with Disorder: Design, Discourse and Disaster, Izmir, Turkey

Buchanan R (2001) Human dignity and human rights: thoughts on the principles of human-centred design. Design Issues, 17(3): 35–39

Chamberlain PJ, Bowen SJ (2006) Designers' use of the artefact in human-centred design. In: Clarkson P, Langdon P, Robinson P (eds.) Designing accessible technology. Springer, London, UK

Chamberlain P, Roddis J, Press M (1999) Good vibrations: a case study of design-led collaborative new product development in the field of vibro-sound therapy. In: Proceedings of Design Cultures: An International Conference of Design Research, The European Academy of Design, Sheffield Hallam University, UK

Dunne A, Raby F (2001) Design noir: the secret life of electronic objects. Berkhauser, Berlin, Germany

Gilmore J, Pine BJ (eds.) (2000) Markets of one: creating customer- unique value through mass-customisation. Harvard Business School Press, Boston, MA, US

Ladyman S (2005) Parliamentary under Secretary of State for Community. Speech given on 1 March 2005: technology and delivery of care for older people speech transcript. Available at: www.dh.gov.uk/en/News/Speeches/Speecheslist/DH_4105328 (Accessed on 14 November 2007)

McCarthy J, Wright PC (2004) Technology as experience. MIT Press, Boston, MA, US

McDonagh-Philip D, Lebbon C (2000) The emotional domain in product design. The Design Journal, 3(1): 31–43

Newell AF (2003) Inclusive design or assistive technology. In: Clarkson J, Coleman R, Keates S, Lebbon C (eds.) Inclusive design: design for the whole population. Springer, London, UK

Norman DA (1990) The design of everyday things. Doubleday/Currency, New York, NY, US

Rust C (2004) Design enquiry: tacit knowledge and invention in science. Design Issues, 20(4): 76–85

Rust C (2007) Unstated contributions – how artistic inquiry can inform interdisciplinary research. International Journal of Design, 1(3)

Sengers P, Boehner K, David S, Kaye JJ (2005) Reflective design. In: Proceedings of the 4th Decennial Conference on Critical Computing (CC'05), Aarhus, Denmark

Sheffield Partnerships for Older People Projects (2006) Expert Elders Network 'our time has come' promotional leaflet, document reference NCC1602, Sheffield City Council, UK

SPARC (2007) Awards – Projects we support – More information on Call 2. Available at: www.sparc.ac.uk/awards_projects_Call2.asp (Accessed on 12 November 2007)

Walters P, Chamberlain P, Press M (2003) In touch: an investigation of the benefits of tactile cues in safety-critical product applications. In: Proceedings of the 5th European Academy of Design Conference, Barcelona, Spain

Walters P, Chamberlain P, Press M, Tomes A (2004) Designing by numbers? Keeping the human in human-centred design. In: Proceedings of Pixel Raiders 2, Sheffield Hallam University, Sheffield, UK

Walters P (2006) Knowledge in the making: prototyping and human-centred design practice. PhD-thesis, Sheffield Hallam University, Sheffield, UK

Z33 (2007) Designing critical design. Exhibition at Hasselt, Belgium, on 4 March – 3 June 2007. Available at: www.z33.be/index.asp?page=detailproject%7Csub=67&lang=en (Accessed on 12 November 2007)

Chapter 5

Biomechanical Analysis of Opening Glass Jars: Using Kinematics

J.R. Fair, L. Bix and T.R. Bush

5.1 Introduction

The research theorizes that there are three styles of hand movement when removing a lug closure from a glass jar; lid hand movement alone, jar hand movement alone, or the movement of both hands. However, it is hypothesized that experimental set up will significantly impact how participants move as they remove jar lids; that an unrestrained subject will move differently from one restrained for ease of experimental set up. Each study participant will open glass jars with lug style closures in two fashions: restrained, so that the test jar remains on the table, and unrestrained.

In order to accurately capture the movement when subjects are not restrained, the system must be fixtured, creating a "reference point." Ten to twenty subjects without a history of injury to the hands, wrists, arms, or shoulders will be used in the study in order to create the fixturing system, refine the methodology and test the hypothesis that restraining subjects influences movement.

Having trouble opening packaging is by no means a new phenomenon (DTI, 2000; Langley *et al.*, 2005). From a packaging perspective, most research that looks at the trouble associated with opening glass jars has been done in relation to the torque applied to the closure (Crawford and Wanibe *et al.*, 2002; Voorbij and Steenbekkers, 2002; Langley *et al.,* 2005; Yoxall *et al.,* 2006). However, torque is a single factor of many that impact utility, or lack thereof, for packaging. Factors such as dexterity, hand size, handedness, posture, frictional forces and motion all contribute to the success, or failure, of the package. The objectives of this research are to:

1. develop two methods that can be used to examine motion when testing packaging (one restrained, one not);

2. determine whether or not restraining the subject significantly impacts movement, thereby interfering with the test results.

Research into this area is needed because: 1) pilot studies conducted by the team indicate that restraining subjects may affect their motion 2) only limited research into biomechanics and packaging is available, and what there is frequently restrains subjects. Research on either motion or torque involving jars, that restricts subject motion, includes Fowler and Nicol (2001), Stins *et al.* (2001) and Murgia *et al.* (2004). Research that allowed subjects to interact with packaging the way they normally would includes Yoxall *et al.* (2006, 2007) along with Voorbij and Steenbekkers (2002). However, Yoxall *et. al.* (2006) and Voorbij and Steenbekkers (2002) focused on the forces involved in removing lids and did little with regard to the movement of the user.

A commonly used industry practice is to apply a lid at a torque equal to one half the finish diameter of the lid (Soroka, 2002). For example, a jar with an 82mm finish would be torqued to 41 in – lbs, or 4.63Nm. This industry "rule of thumb" presents concerns in the light of the available usability research. Voorbij and Steenbekkers (2002) indicate that opening torque should be limited to 2Nm for maximum usability. Their team notes the lack of enthusiasm from packaging professionals for this finding, commenting, "It is strange, therefore, that the packaging industry has not yet addressed the problem by changing this type of packaging" (Voorbij and Steenbekkers, 2002).

Limited research is available regarding what does enable consumers to use jars effectively, and what has been done generally restrains people, not accurately replicating every day use. These restraints alter the way subjects' perform the opening procedure and may or may not provide value for designers. Fowler and Nicol (2001) assessed both the force and motion of opening after looking at anthropometric data of subjects. The study design did not accurately replicate reality in that their jar was not made of glass, and subjects were required to keep the jar on the table. Murgia *et al.* (2004) looked at wrist kinematics during jar opening. However, the subjects were seated and required to open the jar closure with their dominant hand. Voorbij and Steenbekkers (2002) tested over 700 subjects and found that a majority of people prefer to have their dominant hand on the jar, not the lid. However, the same research team did not investigate factors that explained the positioning of the hand on the jar or lid; perhaps handedness, laterality, postural preferences, injury or just random placement determine hand placement and motion.

Mohr *et al.* (2003, 2006) examined postural preferences using arm – folding and hand – clasping. Their research identified four possible groups for these two behaviours: people who congruently place the same side on top (*e.g.* left thumb on top for hand clasping and left arm on top for arm folding), or people who are incongruent (*e.g.* left thumb on top for hand clasping but right arm for arm folding). They found that the greatest number of people prefer their left hand on top in both activities, but this is not related to their dominant hand. For this study researchers will examine data for possible correlations between postural preference and hand positioning.

The physical set up of the experiment is also crucial. Stins *et al.* (2001) noted that the contents of a glass jar and its location, relative to the subjects' hands, were significant factors in how they reached for it. This research will look at the difference in opening styles associated with a jar located on a table, a constrained position, and a jar located in air. Perhaps people open items differently in an x, y, z coordinate system similarly to the x – y system observed by Stins *et al.* (2001).

Unlike most studies related to this topic, Voorbij and Steenbekkers (2002) allowed the subjects to replicate opening in a less controlled environment. However, their research examined the forces associated with the use of an aluminium jar. Similar research conducted by Yoxall *et al.* (2006) built on that of Voorbij and Steenbekkers (2002) but employed a glass jar.

For the purposes of this research, we will explore how people open lug style closures from a motion perspective when constrained to a table surface and when opening in air. In order to make sure all air openings are performed in a consistent manner, subjects will be instructed to begin opening from the position demonstrated upon arrival (see section 5.2.3 and Figure 5.1.). Hand size, strength, postural preferences, handedness, and gender will also be examined and factored into the results.

The ultimate research goal is the development of a methodology that captures the motion of users of varying ability as they open packages so that the information can then be used to inform design.

5.2 Methodology

5.2.1 Subjects

The research will be collected at the Biomechanical Design Research Laboratory (BDRL) in the department of mechanical engineering at Michigan State University. A convenience sample consisting of 10 – 20 students will be used to develop the initial methodology. Subjects that have a history of injury to the hand, wrist, arm or shoulder will be eliminated prior to testing. After the methodology has been established, a small group of 5 – 10 older consumers aged 60 and over will be tested to examine the robustness of the methodology and the effect of aging on motion.

5.2.2 Hand Size, Strength and Postural Preference

Researchers will collect demographic information, including age, gender and preferred dexterity.

Following the collection of the demographic data, a series of digital photos will be taken of the subjects' hands in order to characterize their anthropometrics. These photos will be taken following the procedure set forth in Chapter 6 Finger/thumb dimensions (DTI, 2001).

To quantify hand strength, researchers will collect grip strength, wrist strength, and bilateral palm to palm squeeze strength using a Jamar ® grip dynamometer, Baseline® pneumatic wrist dynamometer, and a Baseline® squeeze dynamometer, respectively. Subjects will be positioned in accordance with the American Society of Hand Therapists (ASHT) standard for all testing employing the dynamometers.

All measurements collected will be used to look for possible correlations with positioning and motion that can be used to inform packaging design to improve accessibility.

Finally, following the "postural preference" procedures established by Mohr *et al.* (2003, 2006), digital photos will be taken of the subjects to determine each subject's postural preference group. This data will be examined for correlation with hand positioning on the jars.

5.2.3 Biomechanical Capture

After the collection of the anthropometric data, subjects will be asked to remove any material that might reflect light (watches, rings, *etc.*) The subjects will be asked to stand at a table set to a counter height of 91.44cm. This height is considered to be the standard work surface height for universal design kitchens (GEAPPLIANCES, 2007).

The testing will use two jars and one lid style. The lug closure will be a regular twist – off style, an RTP. Jar A has a finish designation of 7 – 20.3, a height of 15,082cm, and a diameter of 8,732cm. The lid diameter of Jar A is 7cm. Jar B has a finish designation of 8.2 – 20.4, height of 15,516cm, and a diameter of 8,572cm. The lid diameter of Jar B is 8.2cm. Jars will be filled to two-thirds of their overflow capacity with popcorn kernels to replicate a real life situation.

Because the pilot work has indicated that people grip the jars in differing ways (see Figure 5.1), subjects will be handed a normal jar and asked to open it. The grip that is used by the subject will determine the instrumentation that they receive once testing using the camera system begins (see Figure 5.3).

Figure 5.1. The two opening positions noticed during pilot studies

Kinematical data will be collected using five infrared cameras and Qualisys software with a capture time of four seconds at 60 Hz. A sixth, digital camcorder will be used to capture a video clip of the testing. Infrared light sources mounted around the camera lenses shine onto retro-reflective targets. The targets are spheres of 12.5 mm diameter covered in a reflective tape. The tape has small glass spheres embedded in it that reflect the infrared light. There are three targets on each hand. The first target is placed on phalange II at the joint between the middle phalanx and proximal phalanx, the second target is located on phalange II just above the wrist at the base of the hand, the last target is placed on phalange V at the joint between the proximal phalanx and the metacarpals. An additional target is placed on the right arm; this target serves as an identifier of the right hand and will not be analyzed for position. The target set up is demonstrated in Figure 5.2.

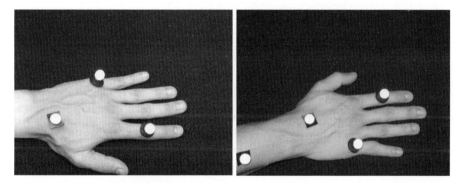

Figure 5.2. Left hand and right had target placement for motion capture. Note that the right hand has an additional target for indication purposes only in the Qualisys program.

Subjects will be tested in a restrained fashion, required to open on a table, and an unrestrained fashion, opening in air, to examine the impact of conducting research in which subjects are required to open on a table surface. However, starting and ending position in air will be set by the research team instructing the subjects to start at the opening position and keep the lid in hand once free of the jar. This will allow the research team to explore differences between the two positions.

Restrained
Prior to testing, the jar will be positioned into a 0.635 cm board that has a hole with a diameter just slightly larger than the jar. This fixturing will provide a consistent positioning of the jar, relative to the cameras and subject. A target will be placed in the centre of the lug closure. The closure will be placed on the jar with a reflective target fixed in the centre and will be captured in the software as a reference for the height and centre of the jar. The target will then be removed from the closure. The subject will be asked to open the jar without lifting it off the table. They will also be instructed to keep the lid on the jar. In order for the lid to be disengaged from jar x the subjects must turn the closure or jar a total of at least 30 degrees. Each of the two jar sizes will be opened twice per subject in this fashion.

Unrestrained

Two rigs will be created to test the jars in an unrestrained fashion. These fixtures are based on the two observed opening styles (see Figure 5.1). The first rig, created to accommodate users who grip the walls of the jar as they open, will have a rod fixed to the base of the jar; this rig is seen on the left hand side of Figure 1.3. The rod has two spheres that create the centre line of the jar, or the axis of rotation, which can then be identified in the Qualisys software. The second rig, created for users who grip the bottom of the jar during the opening process (see Figure 5.1), has spheres located on each side of the jar, again providing positional reference so that relationships between the jar and the hand can be drawn. The side is determined by the line left on the glass from the mould during manufacturing. Both jar rigs are shown in Figure 5.3.

Figure 5.3. The jar on the left is a set up for subjects who place their hand on the side of the jar: the jar on the right is set up for subjects who place their hand on the bottom of the jar

Subjects will be asked to open the jar rig that represents the opening style observed using the same process outlined in the first series.

Data Analysis

Data collected by the Qualisys software will be exported to Excel. Graphs will be created that incorporate the target motion of both the right and left hand relative to the reference lines (see Chapters 5.2.3.1 and 5.2.3.2). A sample graph is shown in Figure 5.4.

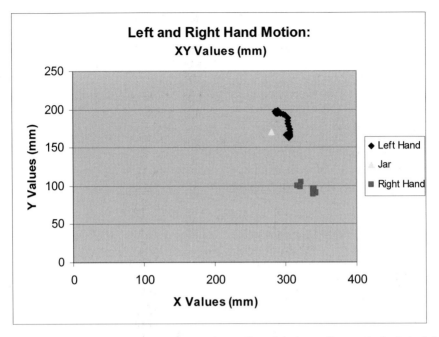

Figure 5.4. This graph shows the x y data points collected during a pilot test for both the left and right hands. The triangle represents the centre of the jar lid allowing vector analysis to be used in order to calculate the rotation of each hand in degrees.

Using the centre of the jar lid as the reference point, vectors can be used to examine which hand supplied the motion to remove the lid from the jar. This could include the lid hand, the body hand, or a combination of both motions. Once the graphs have been created, the research team will decide on limits that will separate the motion into three categories: 1) Lid hand 2) Jar hand or 3) Both hands. The results from the motion capture will be used to determine the differences between the two opening positions. The data collected on demographics, hand strength, and postural preferences (see section 5.2.2) will be examined for possible insights into the reasons for 1) differences between the two opening styles and 2) differences in how people interact with jars. Additionally the data of the older consumer group will be compared with that from the younger group.

5.3 Conclusions

The development of this methodology has significance for both the packaging industry and biomechanical researchers. To date, studies that examine usability and packaging are quite limited. Those that examine motion and packaging are even fewer (Fowler and Nicol, 2001; Stins *et al.*, 2001; Murgia *et al.,* 2004), and the vast majority of these studies significantly restrain people during testing.

There are many factors involved in how consumers interact with packaging. It is paramount that all of these factors be considered and understood in order to design packaging that is truly universally accessible. By building on the existing research surrounding the torque involved in opening glass jars, packaging professionals and designers can begin to understand and create packaging that takes advantage of the way a consumer interacts with the package. By starting with the basics of motion capture we can begin to understand the human package interface. Using this knowledge of how people hold jars design can evolve in a way that allows the consumer to apply a greater force without being aware that anything has changed. If packaging professionals can understand the reasons why consumers interact with packaging in the way they do these new principles can be applied to all areas of packaging development and design.

5.4 Acknowledgments

The authors wish to acknowledge the generous donation of glass jars and closures by St.-Gobain Containers, and Silgan White Cap Americas. Additional acknowledgements are due to Rick Bayer, who coordinated the material donation and has provided advice on industry standards. We also acknowledge the partial funding provided through the Center for Food and Pharmaceutical Packaging Research which provided travel funding and subject incentives.

5.5 References

Crawford JO, Wanibe E, Nayak L (2002) The interaction between lid diameter, height and shape on wrist torque exertion in younger and older adults. Ergonomics, 45: 922–933

DTI (2000) A study of the difficulties that disabled people have when using everyday consumer products. URN 00/1070, Department of Trade and Industry, London, UK

DTI (2001) Specific anthropometric and strength data for people with dexterity disability. URN 02/743, Department of Trade and Industry, London, UK

Fowler NK, Nicol AC (2001) Functional and biomechanical assessment of the normal and rheumatoid hand. Clinical Biomechanics, 16: 660–666

GEAPPLIANCES (2007) Universal design frequently asked questions. Available at: www.geappliances.com/design_center/universal_design/faq.htm (Accessed on 29 August 2007)

Langley J, Janson R, Wearn J, Yoxall A (2005) 'Inclusive' design for containers: improving openability. Packaging Technology and Science, 18: 285–293

Mohr C, Thut G, Landis T, Brugger P (2003) Hands, arms, and minds: interactions between posture and thought. Journal of Clinical and Experimental Neuropsychology, 25(7): 1000–1010

Mohr C, Thut G, Landis T, Brugger P (2006) Arm folding, hand clasping, and Luria's concept of "latent left-handedness". Laterality, 11: 15–32

Murgia A, Kyberd PJ, Chappell PH, Light CM (2004) Marker placement to describe the wrist movements during activites of daily living in cyclical tasks. Clinical Biomechanics, 19: 248–254

Soroka W (2002) Closures. In: Soroka W (ed.) Fundamentals of packaging technology, 3rd edn. IoPP, Naperville, IL, US

Stins JF, Kadar EE, Costall A (2001) A kinematic analysis of hand selection in a reaching task. Laterality, 6: 347–367

Voorbij AIM, Steenbekkers LPA (2002) The twisting force of aged consumers when opening a jar. Applied Ergonomics, 33: 105–109

Yoxall A, Janson R, Bradbury SR, Langley J, Wearn J, Hayes S (2006) Openability: producing design limits for consumer packaging. Packaging Technology and Science, 19: 219–225

Yoxall A, Luxmoore J, Austin M, Canty L, Margrave KJ, Richardson CJ et al. (2007) Getting to grips with packaging: using ethnography and computer simulation to understand hand-pack interaction. Packaging Technology and Science, 20: 217–229

Chapter 6

Building a Consumer Network to Engage Users with Disabilities

Y.M. Choi, D. Sabata, R. Todd and S. Sprigle

6.1 Introduction and Objectives

Involving end users is critical when designing assistive devices or conducting research on disability. Identifying and recruiting users willing to take part in these activities can be a difficult and time consuming task for designers and researchers who want information from a targeted group. Study samples which can be generalized to the larger population are important to both as well. To help address some of these issues, the Center for Assistive Technology and Environmental Access (CATEA) at the Georgia Institute of Technology created the CATEA Consumer Network (CCN). The CCN is a network of older adults and people with disabilities whose members test new prototypes, products and services in order to improve them through focus groups, field-testing and surveys (www.catea.org/ccn/ccn.php).

The idea behind the CCN was to create an active user community as well as a tool to support the needs of researchers and designers. The CCN offers ongoing activities and information exchanges in order to keep users engaged and interested in participating. It also provides a way for researchers and designers to easily connect with a motivated group of persons with disabilities who are interested in participating in research and design projects.

Many uses of the CCN can be envisioned, including

- the identification of new, emerging, or unmet needs in products and services;
- the collection of feedback about the design of current or potential assistive products and services;
- an understanding of how the market for and use of assistive devices changes over time;
- collecting perceptions of how public policy and legislation impacts assistive technology use.

6.2 Design and Operation of the CCN

The concept of building a consumer network is not new and a number of others exist to meet various needs. Some are intended to support research activities and focus on a specific topic or population such as the WirelessRERC's Consumer Advisory Network (www.wirelessrerc.org/for-consumers) for study of wireless devices and the Age Network (www.agenetwork.phhp.ufl.edu/recruit.html) for studies involving older adults. Others are focused on consumer advocacy and policy change such as Cochrane Consumer Network (www.cochrane.org/consumers/about.htm) which is a consumer run network that studies effective healthcare interventions. Many others are focused on support of members with specific conditions such as the Georgia Mental Health Consumer Network (www.gmhcn.org). The design of the CCN extends the ideas behind these efforts to create a unique network that

1. includes a broad member base to support research into many areas of disability;
2. collects demographic data consistent with existing national databases;
3. is designed to regularly engage its members to foster a research-support community.

The CCN was designed to be a research and design assistance tool. As such, it had to be populated by a wide range of persons with disabilities. To better track how CCN membership reflects persons with disabilities in the US, members provide demographic and descriptive information based directly on large surveys such as the US Census (www.census.gov) and the National Health Information Survey on Disability (NHIS-D), (National Center for Health Statistics, 1994), and classification systems such as the International classification of functioning, disability and health (ICF), (www.who.int/classifications/icf/en/).

Registration focuses on six specific areas of disability: vision, hearing, communication, mobility, dexterity and cognition. Information is gathered about the type and severity of the disability, the types of assistive devices that are used as well as basic demographic information and is stored in a secure database. At the end of the registration process, each CCN member is given access to his/her own individual member profile. When members log into the CCN website, they are presented with an interface that allows them to review or update information stored in their profile in the database.

Members are engaged regularly with opportunities to participate in a variety of research and design activities. The frequent activities are meant to help foster a community of end users who can become partners in research and development as well as to reduce the possibility of members losing interest. Members within the local region are can be asked to participate in focus groups and usability studies that require travel to Georgia Tech. Members outside of the region can be engaged to take part in regional or national studies such as surveys or telephone interviews. Currently, short surveys are conducted bi-monthly, so that all members can easily participate. In addition, a quarterly newsletter is also sent to all members that contains the results and updates of ongoing research projects, announcements of upcoming activities and member submitted content. This newsletter and the member-submitted content is always available on the CCN website.

Researchers can search for potential participants through a researcher on the CCN website. An interface page provides a mechanism for authorized researchers to access desired information about the members of the CCN while protecting the identity of individual members. The database can be queried for any demographic or descriptive variable collected from members during registration. This informs researchers about CCN membership with respect to a study's inclusion criteria since the researcher needs to know if the CCN membership can support his or her needs. All of this is done in a private manner so that no personally identifiable information is available to the researcher.

To make contact with CCN members, researchers must submit requests through a gatekeeper. The gatekeeper acts as the interface between the researcher and CCN members since he/she is the only ones with access to the names, addresses and other personal information. The gatekeeper's role is to make the initial contact, on behalf of the researcher, with CCN members that match a study's inclusion criteria. However, before contact can be made, the study and all of its materials must be approved by the Georgia Tech Institutional Review Board (IRB), (www.compliance.gatech.edu/IRB/). The IRB governs all activities that involve the use of the CCN to ensure that all criteria for human subject study are met and that any data collected is handled securely.

6.3 Recruitment and Demographics

New members are recruited primarily through networking with disability organizations, word of mouth and the CCN website. Any person with a disability can become a member by completing the signup survey on the CCN website or via voice/TTY over the telephone. Active recruitment efforts are conducted by reaching out to potential members through disability specific organizations. On average, current recruitment efforts have yielded about 45 new members each month. By comparing the membership of the CCN to national databases (the United States Census 2000 (U.S. Census Bureau, 2004) and the NHIS-D (Adams and Marano, 1995; Maag, 2006)), recruitment efforts can be focused to populate the CCN with a similar demographic mix.

Currently, the 500+ CCN membership has several differences from the population of persons with disabilities in the US. CCN members tend to be younger and have completed a higher level of education than people with disabilities in the general population. About 90% of CCN members are 18–64 years of age and about 9% are 65 years or older. Of all disabled persons nationally, only 64.2% are between 18 and 64 years and 32.4% are 65 and over (see Figure 6.1). This skewed membership is, in part, due to recruitment methods and a remnant of an initial CCN focus on working age adults. In terms of education, 51.8% of CCN members have completed 4 or more years of college compared with only 15.2% of all disabled persons nationally (see Figure 6.2). There are some differences with respect to disability as well. Mobility related disability was identified by 65.8% of CCN members compared with 31.6% of the national disabled population. With respect to hearing, only 22.6% of CCN members a have hearing related disability compared with 51.9% of all disabled persons in the US (see Figure 6.3). Dexterity limitation is a category of information that is collected during CCN registration, however information about dexterity is not easily comparable with national statistics.

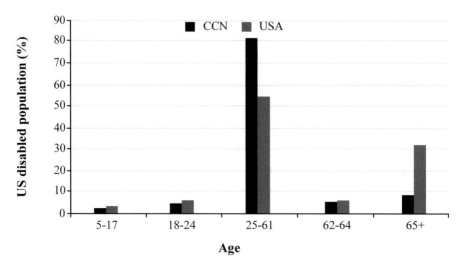

Figure 6.1. Ages of CCN membership compared with those of all persons with disabilities in the US

Knowing the difference between the characteristics of the CCN membership and characteristics of people with disabilities nationally is important for two reasons. As mentioned, it helps focus recruitment efforts to help make the membership more representative of the overall disabled population. Second, researchers can identify biases or limitations in the study samples selected from the CCN which may not be similar to the national representation. At this time the representation in the CCN is largely younger adults, so a study with older adults would be more limited by the current membership particularly when considering additional selection criteria. Understanding these differences in the CCN and national data sets is important in helping to determine the external validity of research conducted using CCN members.

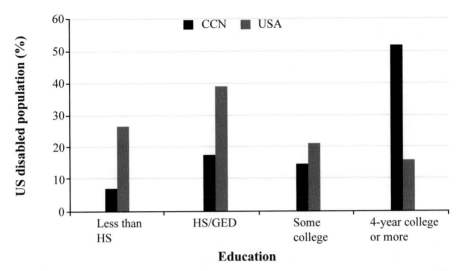

Figure 6.2. Levels of education of CCN membership compared with all persons with disabilities in the US

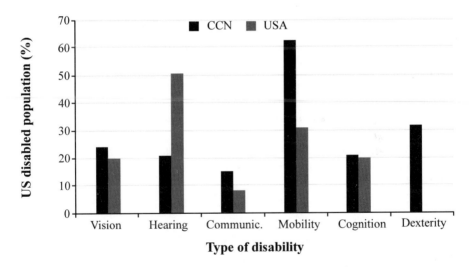

Figure 6.3. The percentage of CCN members with specific types of disabilities compared with all persons with disabilities in the US. *Analysis of national statistics on dexterity is not currently available for comparison.

6.4 Status and Usage of the CCN

Though still new and growing, the CCN is proving to be a useful tool in understanding disability issues directly from an end user perspective. The CCN has been utilized for on-line surveys and as a recruitment tool for in-person studies.

Recruitment has been done for two studies targeting wheelchair users and one targeting workers with disabilities. For recruiting wheelchair users, the CCN membership was queried for members who lived in the Atlanta, Georgia area and who reported using a wheelchair. One study is investigating the lifespan of wheelchair cushions and requires subjects to come to a local hospital for testing. The other study recruited wheelchair users for focus groups administered by colleagues at the University at Buffalo's Rehabilitation Engineering Center on Technology Transfer. The third recruitment effort sought focus group members to discuss accommodations for communication, mobility and visual impairments in the workplace. This study recruited CCN members with visual and mobility impairments for in person focus groups, whereas the group focused on communication was conducted entirely online.

Periodic surveys have also been administered that focus on assessing user needs and AT usage:

- Workplace accommodations. CCN members across the disability spectrum were asked about the types of assistive technology used in the workplace and who paid for it (Zolna, 2007). A follow-up is now planned to track changes over time and to further describe the types of workplace accommodations commonly used today.
- Portable ramps. CCN members who identified as having a mobility impairment were asked about their use and knowledge of portable ramps. The data yielded immediate information regarding ramp use in real world settings, users' perceptions of ramps and opportunities for potential design improvement (Choi, 2007).
- Multiple mobility device use. CCN members with mobility impairments were again asked about mobility device use inside and outside of the home. The purpose was to learn more about persons who use multiple devices and the environments of their use.
- Use of recycled AT. All CCN members were asked questions about used or recycled assistive devices. A recent focus on re-cycling AT motivated us to learn more about it.

The use of short surveys (<20 questions) is made possible by the CCN database structure. Demographic and descriptive information about their functional limitations is entered once only during member registration. The database links this personal information with survey responses via a unique member identification number. This makes all data previously gathered from the member available for analysis. Surveys can thus be kept short and focused specifically on the topic which they are investigating.

6.5 Known Issues and Plans for the Future

Use of the CCN has identified a few opportunities for improved operation in the area of usability, survey response rate, and researcher interface. An ongoing challenge for the CCN is ensuring data security while providing members and researchers sufficient access. Members are required to log into the CCN site to answer surveys and to review their personal information. Currently, email addresses are used as the log-in, but that can prove problematic. Entering an unknown email address blocks log-in. This has happened with users who have changed email addresses and those that have multiple addresses. Blocked access does not allow surveys to be taken and often results in emails to the CCN administrator for assistance. Solutions to this log-in problem are still being investigated.

Rigorous security measures have also been imposed to ensure member privacy. Research privacy policy dictates that personal identifying information be blocked to all researchers except the CCN gatekeeper. However, researchers need access to the de-identified demographic and descriptive information. The current CCN interface currently offers only minimal query capability. Researchers can perform a simple search of one demographic variable to obtain a breakdown of CCN members. For example, a researcher can easily determine the ages of CCN users. However, this current interface does not permit queries using multiple variables. For example, a query for members who are "Male AND users of Manual OR Power wheelchairs" would be helpful in certain applications. A more advanced interface is being developed to permit this added functionality.

In theory, surveying CCN members should elicit a high response rate. By joining, members have consented to participate in research activities and, by extension, should be more inclined to respond to inquiry. However, to date, the response rate has not exceeded 25% of the targeted members. While this is higher than unsolicited surveys, the goal is for a much higher response rate. Our hope is that the quarterly newsletter will help keep members engaged and interested in CCN activities. Because the CCN has now been used multiple times, the newsletter will include the results of prior surveys. We hope that providing members with this information will encourage them to participate. Additionally, a section of the newsletter will be reserved for submissions by members, to convey the message that members are an important part of the CCN and their thoughts are valuable to the CCN community.

6.6 Conclusion

The CATEA Consumer Network was designed as a tool for researchers and designers to easily engage people with disabilities. While consumer networks are not new, the CCN incorporates features that form a unique network. Linking personal information to subsequent survey responses obviates the need for members to enter information multiple times. This reduces survey length and member burden while permitting a fuller analysis of responses. Regular engagement of CCN members is a part of CCN operation as a means to build a sense of community, which, hopefully, will lead to a more vibrant and productive exchange of ideas. Linking demographic information to existing national datasets allows researchers to judge the generalisability of study results while also prompting CCN administrators to recruit members from groups not fully represented in the CCN. As time goes on and more data is collected through future studies, we expect that the CCN database itself will become increasingly valuable in analyzing changing needs and trends in assistive technology.

6.7 References

Adams P, Marano M (1995) Current estimates from the national health interview survey, 1994. National Center for Health Statistics. Vital Health Stat 10(193). Available at: www.cdc.gov/nchs/data/series/sr_10/sr10_193acc.pdf (Accessed on 14 August 2007)

Choi YM (2007) Portable ramp usage of wheeled mobility users. In: Proceedings of RESNA 2007, Phoenix, AZ, US

Maag E (2006) A guide to disability statistics from the national health interview survey - disability supplement. Available at: www.urban.org/UploadedPDF/411307_disability _stats.pdf (Accessed on 16 August 2007)

National Center for Health Statistics (1994) National health interview survey on disability (NHIS-D). Available at: www.cdc.gov/nchs/about/major/nhis_dis/nhis_dis.htm (Accessed on 11 August 2007)

U.S. Census Bureau (2004) Disability status of the civilian noninstitutionalized population by sex and selected characteristics for the United States and Puerto Rico: 2000. Census 2000 PHC-T-32. Internet release date: 14 April 2004. Available at: www.census.gov/ population/cen2000/phc-t32/tab01-US.pdf (Accessed on 13 August 2007)

Zolna J (2007) Perspectives of consumers with communication, vision and mobility impairment on the use and procurement of assistive technology in the workplace. In: Proceedings of RESNA 2007, Phoenix, AZ, US

Part II

Inclusive Design

Chapter 7

Help or Hindrance: The Use of Tools for Opening Packaging

A. Yoxall, J. Langley, J. Luxmoore, R. Janson, J.C. Taylor and J. Rowson

7.1 Introduction

Packaging has to maintain several key functions: preserve and protect the product, make the product appeal to the consumer and not least allow access by the consumer to the contents. Ease of access or 'openability' is becoming a more serious issue for packaging designers, manufacturers and engineers because of the way in which society is changing with huge increases in the proportion of older people. Ageing brings with it many issues, such as loss of strength and dexterity, which can have a major effect on the way in which people interact with everyday items such as packaging. In this paper, the authors undertake an assessment of some of the common tools that have been designed to aid older people in opening common packaging items. The study indicates that whilst some tools are effective, most offer little or no benefit due to the tools themselves not overcoming issues such as loss of dexterity and strength.

Many everyday situations, whether opening a door, boarding a bus or a train, using a phone or getting money from a cash machine can be a difficult task for the older people or people with disabilities. Another everyday item, packaging, whether for food, medicines or other products, has also been found to cause difficulties in terms of accessibility for the older or disabled consumer. A survey of 2,000 people over the age of 50 by Yours Magazine (McConnell, 2004) found that 91% of respondents had had to ask for help in opening a package whilst 71% of respondents had injured themselves trying to open packaging. Opening packaging is, therefore, a huge problem for the older people and those with disabilities. The problem is compounded by the fact that the number of 'older' people in the UK is steadily increasing. By the year 2020, half of the adult population is predicted to be over the age of 50 (National Statistics Office, 2006).

The Yours Magazine survey stated that three out of five people over the age of 50 have purchased tools to help them open packaging. Tools to aid the opening of jars were the second most popular packaging opening aid. However, somewhat alarmingly, a large number of respondents still resorted to using *pliers, scissors, rubber gloves, knives or other tools to try and prise open products*. Clearly, the use of such implements has the potential to cause serious accidents, especially in the hands of the weakest and least dextrous in society.

7.2 Previous Work

There is a huge potential market for jar opening tools to help older people in particular to access safely the jars they buy. There are several jar opening products on the market but an analysis of their effectiveness has never before been undertaken. There have been several previous studies looking at the hand grip strength of consumers. These studies generally employ the use of grip dynamometers (such as the work by Giampaoli et al., 1999), which measures the grip force that a subject can apply across two parallel bars in the palm of the hand. Other such studies use pinch gauges, again measuring strengths not directly related to opening jars or bottles (Mathiowetz et al., 1985). As demonstrated by Crawford et al. (2002) and Yoxall et al. (2006), strength measurements are very dependent on the geometry of the test. This test data is therefore of little use in an analysis of packaging openability, as the forces recorded will not be the same as those the users apply when opening containers. There have also been several previous studies looking at the forces users can apply to jars, such as those by Imrhan and Loo (1988), Rohles et al. (1983) and Voorbij and Steenbekkers (2002). These reports used devices designed to measure the force a user can apply to a jar lid. However, the tests were not fully representative of the real opening situation. Both Imrhan and Rohles fixed the lid to a table rather than allowing the tester to be held in the hand in the same way in which packaging is actually opened. Voorbij and Steenbekkers' tester could be freely held but altered the materials used from those found in packaging altering the friction characteristics and hence the torque measured. This was highlighted in work by Yoxall and Janson (2008), in which torque measurements were taken using a measuring device manufactured from actual items of packaging. Crawford et al. (2002) investigated the way in which lid size and shape affected openability, and discovered that the diameter of a lid affects the torque a user can apply. They also found users could apply more torque to square lids than round lids. None of this previous work involved any tool or other aid to alter the applied torque, and all the subjects used only their bare hands.

This study is not concerned with the actual strength of specific individuals or the average strength of a population as considered in much of the work listed above. Instead, the work is aimed at finding out whether or not using a tool is likely to improve a user's chance of opening a package. The forces that an individual can generate are not considered, only the ratio between the opening forces they can apply just using their hands and the forces using a range of opening tools. This normalised torque reading can therefore be used to rate the tools in terms of the actual increase they are likely to make, regardless of the strength or ability of the subject.

7.3 Aims and Objectives

The main function of a jar opener is to allow a person to open a jar more easily than they would without the tool. Ideally, use of the tool would allow people to open jars that they would otherwise be unable to open. In this way, a tool should allow a greater proportion of the population to be independent, able to perform everyday tasks in the kitchen without requiring help. As the primary market for opening tools is therefore those without the required strength to open packaging without tools, the design of the tools must take into account the strength and ability of these intended users. The effectiveness of tools can therefore be tested according to these requirements by evaluating their ease of use and effect on appliable forces when used by groups of older or infirm subjects. There are a large number of different opening tools on the market. However, many of the devices use identical mechanisms for opening the jar. It was decided therefore to test one example of each type of tool. This should give a good indication of how similar tools will perform. They have been categorized according to whether one or two hands are required for operation. The designs were then further segregated on the basis of the method required to apply the tool to the jar and the opening method. The vast majority of the two-hand devices (where one hand is used for support) use a 'loop' or 'lasso' mechanism, a 'teeth gripping' mechanism or a 'squeeze and maintain grip' mechanism to grip jar lids. All but two of the devices also rely on a lever to provide a mechanical advantage when applying the opening force. The two tools without a lever simply rely on the user twisting the jar lid axially. There are fewer devices available that only require one hand for operation, and all of them grip either the base or lid of the jar, requiring the user to turn the free portion of the package (the jar if the package is held by the lid, and vice versa) to open. Eight tools were selected for testing (shown in Figure 7.1 on page 68) to give one example of each category of tools (except for the 'lasso-extended lever' type tool category, from which two tools with different lasso material were selected).

7.4 Apparatus and Method

The equipment used to measure the opening torques that could be applied with the various tools was custom built and is shown in Figure 7.2 (on page 69). The measurement device consists of an instrumented jar with an embedded torque sensor. The sensor is coupled to a computer via a millivolt sampler so that the torques applied can be recorded. The device was constructed from 'off the shelf' glass jars and lids, and so is able to measure the exact forces that can be applied to a real product during opening as described by Yoxall et al. (2006). Several diameters of glass jar base and jar lid can be used to investigate the performance of the tools with a range of sizes of jar. The various sizes of jar lid are identical except for the diameter, so as to minimise effects due to different closure heights, materials etc. For the study, volunteers were tested individually to avoid any external influence on each individual's responses. They were asked to attempt to

Boa

Undo-it

Zyliss strongboy

Twister

Spill-not

U-grip

Magic twist

Under-shelf opener

Figure 7.1. Various opening tools used in the testing

open the jar testing device as if it were a real jar, without using tools. No restrictions were placed on posture – they were allowed to stand or sit for the test and could hold the jar in any orientation. They were also allowed to rest the jar on a table if they wished. Each subject was then asked to apply 'maximum comfortable force'. It was stressed that the subject should stop the test immediately on feeling any pain or discomfort. This then provided a benchmark for the volunteer's opening ability when completely un-aided. After this first test, a tool was selected and its use demonstrated to the subject, who was asked to repeat the torque measuring test using the selected tool, again stopping if any discomfort was felt. The subjects were given a break between tests to minimise the effect of fatigue. The tools were also tested in random order to prevent any possible effects due to the order of the tests. The tools were also tested on various diameters of jars to see how size affected the performance. Constraints on test time meant that some volunteers did not test all eight tools. Torque data was then used to investigate whether the tools increased the opening torque that a subject could apply from the un-aided level. The dominant hand, tool used and any additional relevant comments were noted for each test.

Figure 7.2. Torque measuring device as developed by the Engineered Packaging Group

Initially, a control group of fully able volunteers was used to assess the tool performance. Testing people who had no dexterity or strength problems showed the operation of the tools when used as per the manufacturers' instructions. Each volunteer was simply asked if they liked the tool they were testing, if they were neutral or if they disliked the tool. A comparison could then be made between the results noted from the control and those from an older people's group, to see if having reduced strength or dexterity affected the performance of the tools. A total of 18 younger subjects, aged between 21 and 25, and 64 older subjects were tested. The older volunteers were aged from 66 to 94 years, with a mean age of 82.

7.5 Results

From the tests, the mean measured torque achieved by older women when un-aided was around 2.2Nm whilst older males averaged 3.7Nm. There was, as expected, a 25% drop in un-aided opening torque from the 65-80 to the 81+ age group but the change in tool performance across the group was not significant. The mean un-aided torque of the younger subject group was 6Nm.

Some of the tools could not operate at all three sizes and so some of the results were limited to just one or two jars. As it is impossible to ensure all volunteers had exactly the same hand size or exactly the same opening method, it was necessary to test several volunteers of both genders and take an average of the tool performance. Figure 7.3 shows the normalised results for the younger control group (a normalised torque value of one indicating that the user can achieve the same opening torque with their bare hands as with a tool). Overall, the best performing tools tested with the younger subject group were the 'boa', 'twister', 'u-grip' and 'spill-not'. These tools were selected for use with the older people's group to see if they still performed well as the user's age increased.

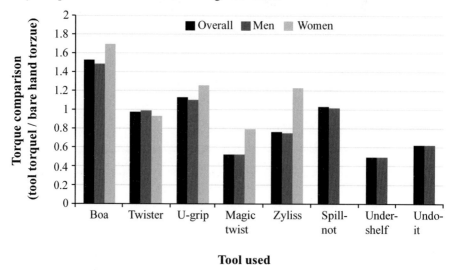

Figure 7.3. shows how normalised torque varies between tools for the younger subjects

The main observation from the results of the tests on the older people was that there was no obvious benefit to the majority of the volunteers using these tools to aid with opening jars. Previous work by Yoxall *et al.* (2006) showed that an average of 3.7Nm is needed to open a typical 75mm jar. The average torque applied by the female subjects when using the tools was around 2Nm or below. The highest average torque produced actually came from the use of bare hands alone (2.41Nm). The boa was shown to be quite ineffective (it had a torque comparison of around one) whilst the u-grip and twister resulted in users applying average opening torques that were lower than when no tool was used at all. The spill-not on the other hand gave an average normalised torque of slightly greater than one, and was found to be particularly helpful by the older people who had no specific medical conditions affecting their hand strength (Figure 7.4). The sturdy base and strong jar support allowed these subjects to concentrate on trying to open the jar rather than worrying about holding it correctly and preventing it from falling.

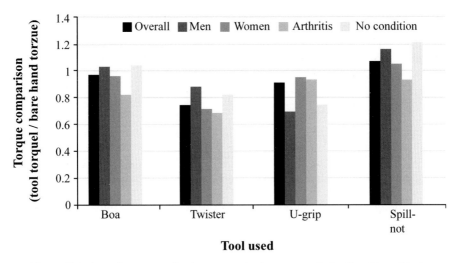

Figure 7.4. shows how normalised torque varies between tools for the older subjects

Whilst in general the tools were shown to perform quite poorly (Figure 7.4), there was a marked difference between the ways in which the tools performed when used by arthritis sufferers compared to the performance when used by older people with no medical conditions. The boa, twister and spill-not all performed significantly worse when used by arthritis sufferers due to the limited dexterity and range of movement that these candidates displayed. The u-grip tool, however, performed better than other tools on average for arthritis sufferers (giving a normalised torque of 0.94 compared with 0.77 for those without arthritis). This may be because significant torques can be achieved just by pushing round one of the arms with the palm of the hand, without squeezing the arms together.

When the results for the older subjects were broken down by different jar sizes (Figure 7.5) it was once again shown that tool performance depends strongly on jar diameter. The smallest, 60mm, jar once again resulted in the best tool performance. There was a significant difference between the two smaller jar sizes and the largest jar however. With the large jar, both the boa and u-grip resulted in normalised torque values of below 0.6. The spill-not, however, still produced a value of around one. This drop was perhaps due to the difficulty that the older people had in holding a tool onto such a bulky jar whilst applying an opening torque. The very wide grip needed to operate the u-grip for large jars may also have been a factor. The spill-not performed relatively well for this jar because the support and grip that it offered meant that the bulky jar size had little impact on the opening torque that the users could apply.

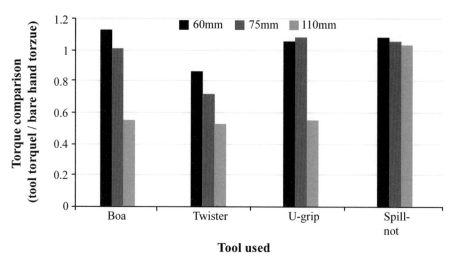

Figure 7.5. shows how normalised torque varies for different jar sizes for the older subjects

Participants were also asked their opinions of the tool and the boa in particular was viewed as extremely unpopular, with around 60% of the volunteers rating it as 'disliked'. The twister was also unpopular, with only 25% of the volunteers rating it as 'liked'. The u-grip got a more positive response from 44% of the people tested rating it as 'liked'; however, around 33% of those questionned also rated it as 'disliked'. The only tool that was generally seen as a good design was the spill-not, liked by 70% of the volunteers. For the older volunteers, the boa offered only a small benefit when attempting to open the smallest jar and performed almost 50% worse than bare hands on the biggest jar. Test observations indicate that this may have been due to a lack of dexterity in the older consumers. Attaching the boa to a jar lid correctly requires several intricate hand movements, and even when correctly in place it can easily slip off the lid. Almost 60% of all subjects (from both age groups) needed some assistance to attach the tool. Once the tool had eventually been attached it was noted that several subjects were unable to apply an opening torque correctly. Many of the older volunteers tended to force the body of the boa tool into the jar lid, deforming it rather than releasing it.

The performance of the u-grip in the tests on older people also suffered because of lack of dexterity. Around 25% of the volunteers were unable to attach it to a jar lid correctly. Even when properly attached, its use required a tight grip across its handles. The handles were relatively far apart, causing problems for many subjects and resulting in many low torque measurements. Maintaining grip using this hand orientation was especially difficult for women (who have smaller hands on average) and arthritis sufferers.

The twister required the user to force down and squeeze tightly during the opening process. This was found to be a problem for anyone with any weakness in their hands or wrists. Testers of all ages complained that they could not use the 'Twister' properly, purely because it 'never seemed to grip the jar'.

When comparing the tool results for the 110mm jar it was seen that although both age groups gained less benefit from the tools than with the smaller jars, the older volunteers performed much worse with the boa and u-grip tools. Observations of the tests indicate this was due to the older volunteers having trouble attaching the tools to such a bulky jar whilst keeping it steady and supported. This was true for both those that suffered arthritis and those that did not. This was much less of a problem for the younger, stronger and more dexterous volunteers. The benefits to an older people of having a jar held securely before trying to open it are shown by the performance of the spill-not tool. Although this tool does not provide any mechanical advantage to the operator, by allowing the user to concentrate solely on opening the jar and not worry about supporting it, the older volunteers were able to apply almost double the force they managed with the other tools for this jar size (Figure 7.5).

It should be noted that there was a significant range in the mean overall performance of each tool. The standard deviation of the normalized torque values for the boa was the largest, at around 0.65 for both age groups. The other tools tested had standard deviation values of 0.5 or below. This level of error was not unexpected as the overall results incorporated data from both men and women (including some subjects suffering from arthritis and similar debilitating conditions). In addition, the results from three different jar sizes were included. The results for the different diameters of jar varied significantly. The overall figures are useful however, because it gives an indication of the overall performance of the tools regardless of the age and health of the operator or the size of jar being tackled. Even when the large deviations in mean normalised torques are considered, the results still indicate that the tools tested were at best marginal in their benefits.

7.6 Conclusions

From all of the tests carried out in the investigation it was clear that the jar opener designs currently on the market are far from perfect. From the initial torque tests on younger, stronger, more dexterous volunteers, it was clear that even when jar opening tools are used in the correct manner, they are often still inferior to using bare hands alone when trying to open a jar. Most jar openers did offer some benefit when applying an opening torque to smaller jars. However with larger jar sizes the benefits from all tools became almost negligible.

From testing volunteers over the age of 65, it was clear that older people experience the same problems as the younger volunteers – even when used correctly the tools offer little advantage. In several cases, the performance of the tools was further hampered by a lack of dexterity or painful hand movements. By comparing tool torques with bare hand torques over the range of jar sizes there was very little evidence to show that jar openers have been designed with the abilities of older people in mind. There was little to suggest that they could offer any benefit, even to those without any medical conditions affecting hand strength. In general, the torque levels recorded suggested that most of the volunteers (especially the women) would have severe problems when trying to open jars, whether they used a tool or not. Some of the tools also suffered the problem that they were very tricky to attach for older people. For example, the boa was rated poorly by many of the older people, although when used correctly it was generally able to apply the highest torques of any of the tools.

This investigation did however reveal that when a jar body is supported and held securely by a tool, then an older person will often be able to apply a greater opening torque, especially if the subject does not suffer from arthritis or similar. The benefits of such a tool were strongly highlighted by the results of the spill-not with the 110mm jar. Hence, future jar opener designs should be much more focused on the abilities of the intended market (*i.e.* older people or those with physical disabilities). They should be easier to attach and simpler to utilize with their use being less dependent on intricate wrist or finger movements.

7.7 References

Crawford JO, Wanibe E, Nayak L (2002) The interaction between lid diameter, height and shape on wrist torque exertion in younger and older adults. Ergonomics, 45(10): 922–933

Giampaoli S, Ferrucci L, Cecchi F, Lo Noce C, Poce A, Dima F *et al.* (1999) Hand-grip strength predicts incident disability in non-disabled older men. Age and Ageing, 28: 283–288

Imrhan SN, Loo CH (1988) Modelling wrist-twisting strength of the elderly. Ergonomics, 31: 1807–1819

Mathiowetz V, Kashman N, Volland G, Weber K, Dowe M, Rogers S (1985) Grip and pinch strength: normative data for adults. American Journal of Physical Medicine and Rehabilitation, 66: 69–72

McConnell V (ed.) (2004) Pack it in! Just say no to impossible packaging. Yours Magazine, 30 January–27 February: 16–18

National Statistics Office (2006) UK Government Actuary population projections. Available at: www.gad.gov.uk/Demography_Data/Population (Accessed on 7 December 2007)

Rohles FH, Moldrup KL, Laviana JE (1983) Opening jars: an anthropometric study of the wrist twisting strength in elderly. In: Proceedings of the 27th Annual Meeting of the Human Factors Society, Norfolk, VA, US

Voorbij AIM, Steenbekkers LPA (2002) The twisting force of aged consumers when opening a jar. Applied Ergonomics, 33(1): 105–109

Yoxall A, Janson R, Bradbury SR, Langley J, Wearn J, Hayes S (2006) Human ability and openability: producing design limits for consumer packaging. Packaging Technology and Science, 19(4): 219–225

Yoxall A, Janson R (2008) Fact or friction: a model for understanding the openability of wide mouth closures. Packaging Technology and Science, (In press)

Chapter 8

The Sound of Inclusion: A Case Study on Acoustic Comfort for All

A. Heylighen, G. Vermeir and M. Rychtáriková

8.1 Introduction

The past decade has witnessed a major shift in how human limitations are viewed (Froyen, 2006). Whereas disability used to be considered as a consequence of physical and/or mental limitations, it is increasingly recognised that the (built) environment holds responsibility for the exclusion of people as well, in other words that the environment can create handicap situations. This shift is complemented by a shift in professional attitudes: design for special needs, which focuses on a specific target group, gradually makes way for universal design, design for all or inclusive design. Inclusive design aims at objects and environments that are accessible, useable and comfortable for full range of people, independently of their capacities and limitations, and this throughout their entire lifespan.

While all people differ in terms of capacities and limitations, human functioning can be generally described by the ergonomic principle (Wijk *et al.*, 2003). This principle assumes that people perceive something in their environment, interpret what has been perceived, act on the basis of this interpretation and remain healthy while performing these three 'ergonomic tasks'. Functioning in the built environment involves at least three modes of perception: seeing, touching and—the focus of this paper—hearing. Hearing provides us with auditory information about our environment, which is vital for human interaction and knowledge transfer, and essential for spatial orientation. Hearing well requires attention to the acoustic qualities of a space and sound amplification. It is especially important for people with a hearing or visual impairment, but also for the growing group of elderly people or for people who have no thorough command of the language of the speaker.

To our knowledge, however, studies on inclusive design in architecture have paid little attention to hearing so far. They tend to focus on accessibility of buildings and spaces, but rarely address their acoustic qualities. The same holds for accessibility regulations. Some prescribe or recommend the instalment of an induction loop, but since this serves only those users who wear a hearing aid, it does not guarantee access, usability and comfort for all. More generally, guidelines exist for background noise level and reverberation time—two factors that considerably impact the acoustics in a space—yet their presence in codes and norms (and in building practice) varies considerably. In some countries, the guidelines are imperative, while others consider them as recommendations (Karabiber and Vallet, 2003). Across the board, however, little attention is paid to the diversity of people's hearing capacities and needs.

The scarcity of attention to acoustic comfort in designing inclusive environments is highly regrettable, because people who do not hear well quickly end up in social isolation as communication becomes difficult. The study reported on in this paper attempts to contribute to the knowledge base by exploring the notion of acoustic comfort for all. The term acoustic comfort is used here to denote a state of mind of satisfaction with the acoustics of the surrounding environment. Just like thermal comfort it differs from one space to another and depends on the activity taking place in that space. For instance, the preferred air temperature in a day care centre differs from that in a beer factory. Similarly, some places require speech privacy (*e.g.* open plan offices), whereas others need speech understanding (*e.g.* auditoria). Furthermore, music typically needs more reverberation than speech. In this paper, the notion of acoustic comfort for all is explored in the context of university education, a context where hearing well is vital to the core activity. After briefly introducing the context, focus and methodology of the study, the paper presents the major findings, followed by a summary and lessons learned.

8.2 Context, Aim and Methodology

8.2.1 A University as a Universal Design Lab

The case study reported on in this paper was conducted within the context of the Katholieke Universiteit Leuven (K.U.Leuven). It is one of the initiatives that use the university as a laboratory for testing mechanisms to foster universal design (Heylighen and Michiels, 2007). Other initiatives include an assessment of the accessibility of the campus by architecture students in collaboration with users/experts; the development of a K.U.Leuven Code, which specifies accessibility norms and guidelines for new build and rebuild projects on campus; or the equipment of K.U.Leuven libraries with reading aids for people with visual impairments.

In an educational context such as a university, universal design has an important pedagogical aspect. Just as a multicultural curriculum is needed to create ethnic and cultural tolerance and diversity, universal design is needed to encourage inclusion and acceptance of all abilities (Tepfer, 2001). Since young people are educated as much by example as through teaching, environments that segregate teach acceptance of segregation and inclusive environments teach inclusion. Unfortunately, many educational environments are rife with examples of non-inclusive settings: "While this offers picturesque examples for UD instructors and powerful experiences of exclusion, it also subtly indicates to students that these characteristics are tolerable aspects of the built environment" (Welch and Jones, 2001).

8.2.2 The Grote Aula

Within the context of the K.U.Leuven, one auditorium was selected to explore the notion of acoustic comfort for all. The 'Grote Aula' (large auditorium) was chosen because of its frequent use for lectures and musical events, the complaints of experts/ users about its poor acoustics, and the plans to renovate it in the near future.

The Grote Aula of the Maria-Theresia college is situated in the old city centre of Leuven. In 1778 Maria-Theresia accommodated the seminar of theology in this former Jesuit college. The right wing in the inner courtyard, which still accommodates the Faculty of Theology, was added in 1786-87. The Grote Aula and 'Kleine Aula' (small auditorium) followed in 1825-27. These neo-classicist auditoria, designed by Martin Hensmans, have a half-round ground plan and are on one floor. Since 1975, the Grote Aula is officially listed as a protected monument, together with the entire Maria-Theresia college and the neighbouring 'Pauscollege' (Pope's College).

At the time of writing the 495 seats of the Grote Aula accommodate undergraduate and graduate students on a daily basis. Once a week, students of the University for the Elderly attend lectures here, and at night, or between semesters, the auditorium is available for musical activities. Both types of events – lectures and musical events – require different acoustic conditions: a speaker demands a much shorter reverberation time than an ensemble. Experts/users especially complain about the difficulty of clearly understanding the speaker during lectures. Comparison with other auditoria at the K.U.Leuven confirmed that the acoustic conditions in the Grote Aula show considerable room for improvement. The planned renovation in 2009 provided extra reasons to submit this auditorium to an in-depth analysis and to propose interventions that improve the acoustic comfort for all.

Figure 8.1a-b. The Grote Aula: exterior (photo: K. De Leye & E. Dooms) and interior (photo: M. Rychtáriková)

8.2.3 Methodology

Inclusive design confronts architects with a new challenge. If we are to design environments that are accessible, useable and comfortable for more people, a human-centred process should replace the design-centred process that characterizes non-inclusive design (Paulsson, 2006). Thus, if we are to provide acoustic comfort for all, we have to give primary attention to the users and their diverse hearing capacities and needs.

To start with, a series of in-depth interviews were conducted with various experts/users – students and personnel with a hearing impairment, students with a visual impairment, and students who attend the University for the Elderly. These interviews provided us with a rough idea of the acoustic comfort in the K.U.Leuven auditoria – which motivated the selection of the Grote Aula for the case study. At the same time, they allowed us to start mapping the hearing needs and preferences of people with diverse capacities and limitations.

Subsequently, an attempt was made to address these needs and preferences in the context of the Grote Aula. The auditorium's present acoustic conditions were measured in situ. In order to improve these conditions, spatial interventions that respect the building's historic character were searched for and simulated. In addition, technical solutions that could further improve the acoustic comfort in the auditorium were inventoried so as to include people with and without a hearing impairment.

The following section will outline the findings of this study without going into too much technical detail.

8.3 Findings

8.3.1 Experts and Users Speaking

In order to get a view of diverse hearing needs and preferences in the context of university education, six students with a hearing or visual impairment were interviewed. These experts/users in the domain of acoustic comfort were contacted via the Working Group Students with Disabilities of the K.U.Leuven and participated voluntarily in the study. In addition, one interview was conducted with representatives of the University for the Elderly. All interviewees were asked to nominate the best and worst auditorium in terms of acoustic comfort, and to indicate problems they experience when attending lectures. The interviews were audio-taped, the tapes were transcribed, and the transcriptions were summarized.

While opinions on the best auditorium varied, several experts/users labelled the Grote Aula as the worst. The strong resonance and long reverberation especially are considered as negatively impacting the acoustic comfort of this auditorium.

Across the board, resonance and reverberation are cited most frequently by all interviewees as causing hearing problems during lectures. Students with a hearing impairment need a clear and high-quality signal in order to be able to understand the speaker, because their remaining sense of hearing is limited. Students with a hearing aid often use an FM listening system during lectures. The speaker speaks into the microphone, after which the speech signal, including the resonant sound, is sent to the hearing aid. According to one student, this is so difficult to understand that she must concentrate entirely on comprehending the speech and cannot make notes. In other words, an FM system as such does not offer a solution; it requires good acoustic conditions to work properly.

Also students with a visual impairment complain about auditoria with long reverberation. In these spaces, they receive so many inputs that orientation becomes extremely difficult. Since they move relying on their sense of hearing (and touch), it becomes almost impossible for them to orient themselves. Moreover, audio-recording lectures is of little use in these auditoria, since the sound quality of the recordings is very low.

Besides reverberation, the experts/users mention several other factors that affect acoustic comfort. Background noise – caused by chatting students, for instance, or creaking wooden seats – is very disturbing, especially for students whose hearing aid does not filter the sounds it amplifies. Other elements include speakers' articulation, their (in)ability to deal with technology, or their tendency not to repeat questions asked in the audience.

8.3.2 Current Situation

As the Grote Aula was identified by several experts/users as worst auditorium in terms of acoustic comfort, we wanted to have a detailed view of its present acoustic conditions. To this end measurements were conducted in situ, which enabled us to derive several acoustic parameters. Based on the results of these measurements, we could assess the auditorium in more detail and propose interventions for improvement. The following paragraphs briefly introduce the reverberation time and the speech intelligibility – the most important acoustic parameters in the context of this paper – without the technical details, and summarize the major results of the measurements and their implications.

Reverberation Time
As mentioned above, long reverberation strongly affects acoustic comfort in a negative way. The acoustic parameter that models this phenomenon is the reverberation time (RT). A long RT (of several seconds) extends syllables so that they overlap, and thus negatively impacts speech intelligibility (DfES, 2003). This is the case in spaces with hard, reflecting walls and ceilings. A long RT causes the background noise level to rise, which further reduces speech intelligibility.

For people with a hearing impairment the allowed RT is shorter. A too long RT increases the background noise and makes listening very tiring and perhaps even impossible. For people with a visual impairment a short RT is crucial too. In a room with a long RT, these people have difficulty locating from which direction the sound is coming. However, the RT has also a minimum limit. Too much absorption in a space, especially in auditoria, diminishes the early sound reflections to such an extent that the sound level produced by the speaker quickly decreases and is too weak at the back (DfES, 2003).

According to our measurements the RT in the Grote Aula is 2.40 seconds, which was to be expected given its size (3,588 m^3) and materials (hard surfaces and leather seats). This is too long – for lectures and for music – and should diminish drastically if we are to provide acoustic comfort for all. In a space of this size, one second would be ideal for speech, while music demands 1.8 seconds. These values are based on experience, since no norm exists.

Speech Intelligibility

Speech intelligibility is an intrinsic characteristic of speech transmission for a certain position of source and receiver. The space (and its shape, finishes, ...) clearly affects the quality of the spoken signal during the transfer from speaker to listener. Reverberation, reflections and concentration of sound waves have a negative effect. Speech intelligibility also depends on the location of the speaker and listener, background noise and the loudness and quality of the spoken signal. Good speech intelligibility implies a strong and understandable signal. The level of occupation plays a role too, since the audience acts as an extra source of absorption, which diminishes the RT and thus improves speech intelligibility.

The speech transition index (STI) of a space varies between zero and one; the higher the value, the better the speech intelligibility. An STI of 0.6 is considered as desirable, yet complete intelligibility of sentences requires 0.75. Research on speech intelligibility with elderly (suffering from reasonable hearing loss) and non-native listeners and speakers, shows an increase in STI of minimum 0.10 to be desirable, yet in many cases 0.15 or 0.20 is necessary.

Judging from the measurements, the STI in the Grote Aula varies between 0.36 and 0.50. Across all positions, the average is 0.44, which is usually experienced as poor. The value diminishes at the back, since the sound level reduces with the distance from the source. The more absorption in the space, the greater this reduction. Moreover, the farther from the speaker, the more reflections (with walls, ceiling) have taken place. If these reflections arrive at the listener with a considerable delay (more than 50 ms), they are disturbing and reduce speech intelligibility.

When using the current audio system in the Grote Aula, the measured STI varies between 0.37 and 0.69, with an average of 0.48. This suggests that the system does not perform properly. A better installation that divides the sound across the entire auditorium would considerably improve the speech intelligibility. The current system makes speech sound louder but not better. The amplified sound, which is broadcast at the front, is strongly reflected by the surrounding walls and ceiling, and causes disturbing echoes. Thus, the audio system does not improve the RT or speech intelligibility.

8.3.3 Improvements

Judging from the measurements in situ, providing acoustic comfort for all in the Grote Aula will require major improvements. A first set of improvements come down to adding absorption so as to reduce reverberation and increase the speech intelligibility. As the auditorium is a protected monument, discrete solutions are needed that can provide extra absorption. In the case of a new building, many more possibilities exist to create a favourable acoustic climate. Given the double function of the Grote Aula – lectures and music – we explored the possibility of applying movable interventions, which can be adapted to the activity taking place.

The following interventions were considered:

1. replacing the leather-covered seats by lightly padded seats covered with cloth;
2. replacing the linoleum floor by carpet (suggested by the technical services);
3. applying an acoustic plaster on an acoustic absorption layer to the wall;
4. placing removable sound-absorbing wall panels covered with cloth;
5. hanging curtains along the wall;
6. applying an acoustic plaster to the ceiling above the gallery;
7. applying a thin acoustic plaster to the cupola and vault above the podium.

Combinations of the interventions proposed for seats, floor, wall and ceiling were tested in an acoustic simulation program with regard to their impact on the spatial acoustics of the auditorium. Table 8.1 schematically presents how they perform on different criteria:

- speech: RT of ca. 1 second and speech intelligibility > 0.6;
- music: RT of ca. 1.8 seconds;
- maintenance: how easily materials can be cleaned and how fast they get dirty;
- flexibility: adaptability to change in use (music/speech);
- replacement: how easily solutions can be removed or replaced.

Note that the score for the latter three applies to the intervention in bold only.

Table 8.1. Performance of interventions. Key: -- = weak, ± = moderate, ++ = optimal

	Present	Interventions								
		1	**1,2**	**1,2,3**	**1,2,4**	**1,2,5**	**1,2, 3,6**	**1,2,3, 6,7**	**1,3**	**1,6**
RT	2.48	2.44	1.72	1.18	1.10	1.26	1.17	1.01	1.90	2.22
STI	0.45	0.46	0.51	0.58	0.58	0.59	0.59	0.62	0.52	0.48
Speech	--	--	--	±	±	±	±	++	--	--
Music	±	±	+	±	±	±	±	±	+	±
Maintenance	++	±	-	++	±	±	++	++	++	++
Flexibility	--	--	--	--	±	++	--	--	--	--
Replacement	±	±	±	--	++	++	--	--	--	--

The RT clearly decreases as absorption increases. Solutions that are labelled as moderate for speech do not meet the requirement for RT and STI, but may come close. Solutions marked as weak are to be avoided, as the RT is way too long and the STI too low. Solutions that do not meet the requirements for music are marked as moderate, because the echo in the space is either too strong or insufficient for music to sound optimally. Cloth materials or covers are moderate solutions with regard to maintenance; they easily absorb dirt that is difficult to remove. Carpet is even worse, as it quickly becomes dirty when walked over. Moreover, it is less comfortable for wheelchair users or people with allergy. Curtains are considered flexible because they can easily be opened or closed depending on the activity taking place (lecture/music). Wall panels can be removed, but this is much more laborious. Plasterwork cannot be removed or replaced without damaging the underlying structure, and is therefore marked as weak with respect to flexibility and replacement. Curtains and wall panels are easily removable, while the moderate solutions require more effort.

Depending on the priorities set, one can select the most appropriate solution from the scheme above. Obviously, the fact that the Grote Aula is protected as historic monument does play a role in this selection. However, whatever solution is selected, increasing absorption alone will not automatically lead to acoustic comfort for all. As the table indicates, none of the solutions perform optimally for reverberation and speech intelligibility – even less so when taking into account the more stringent acoustic conditions posed by people with a hearing or visual impairment. Since these are not attainable in the context of the Grote Aula, the acoustic comfort needs to be further improved in a technical way.

Several technical interventions may improve speech intelligibility in an echoing space. Within the gamut of loudspeakers alone, multiple types and possibilities exist. In this study, we have only considered interventions that can bring the sound signal to the listener with better quality. An induction loop and the use of headsets avoid sound waves having to cover a path through free space before reaching the listener. As a result, the poor acoustic qualities of the space have less impact on the quality of the signal transmitted. Loudspeaker installations solve this problem by directing the sound so that it avoids the reflecting surfaces in the space as much as possible. Moreover, the loudspeakers should be oriented so that the sound level no longer diminishes with the distance to the source, as is currently the case in the Grote Aula. Applying extra absorption will further reduce the sound level in all positions and thus make well-oriented loudspeakers even more important. Unlike the interventions to increase absorption, the proposed technical solutions have not been tested with simulation software because of the laborious and very specific character of the subject matter. If specific performance requirements are formulated in advance, these calculations or simulations can be performed by the suppliers of these installations.

8.4 Summary and Lessons Learned

In architecture, inclusive design aims at environments that are accessible, useable and comfortable for all people. This aim is usually associated with accessibility, yet acoustic qualities may considerably impact usability and comfort as well. By way of first step, the study reported on in this paper explored the notion of acoustic comfort for all in university education, a context where hearing well is vital to the core activity.

The experiences of diverse experts/users revealed that the Grote Aula performs very weakly in terms of acoustic comfort. Measurements in situ confirmed the need for radical improvements. Several interventions regarding material use were proposed—taking into account the auditorium's historic character and double use—and tested in terms of effectiveness via a simulation model. Analysis shows that the ideal solution does not exist; what is most appropriate depends on the priorities set (lectures versus music, maintenance, flexibility, replacement). Since none of these interventions guarantees acoustic comfort for all, including people with a hearing or visual impairment, additional technical interventions are proposed: a better audio system, a headset connection and an induction loop. In addition, speakers' awareness should be raised regarding the importance of articulation, correct technology use, and repeating questions asked in the audience.

The Grote Aula is a historically protected monument and since every monument is protected in a specific way, general guidelines are difficult to derive. Nevertheless, the outcome of the case study largely transcends the planned renovation of this specific auditorium. The issues inventoried and addressed in this study may serve as points to consider when (re)designing other auditoria in the future. Moreover, the solutions explored in the context of the Grote Aula may be useable in other renovation projects. The same holds for the approach of looking for flexible changes by using acoustic predictions. Judging from our experience, room-acoustic simulations provide a convenient method to evaluate particular alternatives in the context of historical buildings. As such, the study contributes to sensitization and knowledge transfer, in that it makes students and professionals in the building sector aware of the importance of acoustic comfort for all, and offers a first view of which solutions are possible and desirable, and how these can be obtained.

8.5 Acknowledgements

The case study reported on is based on the Master thesis of Karolien De Leye and Eva Dooms. Special thanks are due to the experts/users, the Technical services of the K.U.Leuven, and KIDS (Royal Institute for Deaf and Speech Impaired).

8.6 References

DfES (2003) Acoustic design of schools – a design guide. TSO, London, UK
Froyen H (2006) Barrières detecteren tussen mens en plek, bruggen. In: Ontwerpen voor Iedereen. Ministerie van de Vlaamse gemeenschap, Gelijke Kansen, Brussel, Belgium
Heylighen A, Michiels S (2007) A university as universal design laboratory. In: Proceedings of Include 2007, Helen Hemlyn Research Centre, London, UK
Karabiber Z, Vallet M (2003) Classroom acoustics policies – an overview. In: Proceedings of the 5th European Conference on Noise Control (Euronoise 2003), Naples, Italy
Paulsson J (2006) Universal design education. European Institute for Design and Disability Sweden and Neurologiskt Handikappades Riksförbund, Göteborgs Tryckeriet, Sweden
Tepfer F (2001) Educational environments: from compliance to inclusion. In: Preiser WFE, Ostroff E (eds.) Universal design handbook. McGraw-Hill, New York, NY, US

Welch P, Jones S (2001) Advances in universal design education in the United States. In: Preiser WFE, Ostroff E (eds.) Universal design handbook. McGraw-Hill, New York, NY, US

Wijk M, Drenth J, van Ditmarsch M (2003) Handboek voor toegankelijkheid. Reed Business Information, Doetinchem, The Netherlands

Chapter 9

Designing an Inclusive Pill Dispenser

H. Dong and N. Vanns

9.1 User Needs

Inclusive design is becoming topical but to many people the inclusive design process still seems a mystery, especially to design students who do not often have the chance to work with end users.

This paper aims to illustrate a typical inclusive design process using a final year student design project (*i.e.* Major Project) as an example. The project lasted six months and delivered a pill dispenser working prototype at the end.

The project followed a total design method, and was informed by people with Multiple Sclerosis (MS) throughout. It demonstrated how a typical systematic engineering design approach coupled with extreme user input could result in a mainstream product that benefits older people and those with disabilities.

As an excellent Major Project at Brunel University, the project also offers a unique insight into the social, technical and commercial aspects of inclusive design, thus being of interest to a wide audience: students and tutors from art, design, and engineering institutions as well as designers from the commercial world.

A Brunel Design Major Project typically starts with defining the design problem. This often includes identifying the user group, the main difficulties they have where design could play a role to improve the situation, and initial product analysis to identify the market opportunity.

9.1.1 Users with Multiple Sclerosis

When Natalie Vanns started her Major Project, she focused on a specific group of users: Multiple Sclerosis (MS) sufferers. MS is a debilitating disease affecting 85,000 people in the UK alone (www.mssociety.org.uk). It is a condition which develops typically in twenty to forty year old adults, and can cause serious weakness, loss of vision, numbness in limbs, incontinence, cognitive dysfunction and depression. Many sufferers use a wheelchair for mobility and require support on a daily basis. Around 70% of MS sufferers are female. There is no cure for MS, but its many symptoms are treated with medication on an individual basis. As a result, MS sufferers take a wide range of oral medication on a daily basis, and some users take more than 20 different types of medication a day.

A group of ten women with MS were interviewed at the early stage of the project to discuss the everyday problems they encountered. Dispensing medication was brought to the fore as one of the most frustrating tasks as they lack the strength and dexterity required to open blister packs independently. The MS sufferers interviewed took a minimum of five different types of medication each day. Generally all of the medication was taken at once, either last thing at night or first thing in the morning. This is because of the medication was dispensed by a carer and it was therefore more convenient to take it all at once. Although it is possible to have the daily medication dispensed in non-blister packs, this means making specific requirements of the pharmacist and none of the MS users interviewed actually chose to do so.

This early user study confirmed that design exclusion arises when task requirements exceed user capabilities (Clarkson and Keates, 2003). Further observational analysis revealed that the range and variation of capabilities in hands of people with MS were vast. Based on a grip strength testing with five users with strength loss in hands, it was estimated that the maximum value the user should have to exert in order to dispense medication was around 10N. Aiming to develop a product that helps MS sufferers maintain independence when taking medication, Natalie studied the force requirements of dispensing pills from blister packs.

9.1.2 Force Testing of Blister Packs

A range of typical blister packs used by MS sufferers were selected for force testing, including paper-fronted tablets, foil-fronted tablets, and foil-fronted capsules. These packs were tested on an Instron force application measuring system (Figure 9.1)

Figure 9.1. Force testing method

All of the blister packs tested burst at below 30N with the exception of the large tablet Co-codamol packs (paper-fronted lidding, child proof) which required an average breaking force of 53.7N. It was found that blisters located at an extremity (*e.g.* corner), edge or surrounded by burst blisters required more force to break open than those surrounded by unbroken foil.

From the force testing, it was concluded that the product to be developed should exert at least 30N on the pill each time to break the lidding material. Therefore it should amplify the user force by at least a factor of three and medication such as Cocodamol would be accommodated through additional force exerted using the body.

9.1.3 Analysis of Competitor Products and Parallel Products

Competitor products on the market (Table 9.1) and parallel products that perform similar functions to those of hand-held medication dispensers were then analysed.

Table 9.1. Competitor products on the market

Products	Analysis
	Poppet Pill Dispenser: Dispenses small tablets, not suitable for capsules. Poor affordance as a hand-held device, with placement and accuracy problems. Lacks ergonomic design; no contouring on the product to aid grip. Very cheap and easy to manufacture. Low finish quality.
	Pill Poppa: Targets those with visual problems and arthritis by providing support around the blister pack and catching the pill. Not suitable for capsules (central cup too small). Good affordance. Good material. Can be used one-handed but does not reduce the amount of force necessary to dispense a pill.

A range of parallel products were chosen according to their grip type, material, size, action and the force required to operate them. These included traditional and junior Dymo, staple remover, single-hole punch, cherry stoner and OXO Good Grips kitchen tools. They were presented to people with MS for testing in order to obtain insights into certain qualities the final product should have. It was found that the mechanisms which could be operated both in the hand and on a table surface were most promising, as the user can apply body force if required. Clear visibility at the work end was also important.

Based on the user testing, especially the users' preference of form and material, it was concluded that the product to be designed required a small hand span, but to be wide enough to ensure no slipping or movement in the palm. It should utilise elastomer grips on key contact points with the body and have no exposed metal surfaces as points of contact.

9.2 Design Process

With the early insight into the design problem through user and market research, a typical systematic engineering design process started with Product Design Specification, followed by Conceptual Design, Embodiment Design and Detail Design (Pahl and Beitz, 1996).

9.2.1 Preliminary Product Design Specification

The Product Design Specification (PDS), (Table 9.2) aimed to outline the necessary constraints and conclusions highlighted in the research of this project. The categories of the PDS referenced (Pugh, 1991).

Table 9.2. Product design specification

Category	Specification	D/W *
User group	Primary user group: female MS sufferers of 18+	D
	Low reliance upon user strength and dexterity	D
	Secondary user group: older people and those with disabilities, *e.g.* stroke and arthritis sufferers	D
Market size	8,500 (10% of the MS sufferers)	D
	824,500 (including other potential users)	W
Ergonomics	Power grip must be used to achieve max. force output	D
	Wrist should be kept in a neutral position	D
	Hand span must be kept within 45mm-60mm	D
	Other areas of the body should be considered as an alternative for force exertion instead of the hand	D

Category	Specification	D/W *
Ergonomics (cont.)	Must fit comfortably in the hand without slipping	D
	Operation with one hand	W
Environment	Portable – lightweight but durable	W
	Should withstand a fall of 1.5m	D
	Perform between temperatures of 0 and 35 degrees.	D
	Perform between humidity levels of 10-90%	D
Weight and Size	If portable: minimum weight and size	W
	If stationary: consider weight balance, more space for user interface	D
Performance	Dispense capsules of dimensions: 21.5mm x 7.45mm	D
	Dispense tablets of diameters 7mm to 15mm	D
	Should not break the medication when dispensing	D
	Should amplify user force ($\leq$10N) by a factor of 3	D
Service life	Thirty operations each day for five years	D
Maintenance	As low maintenance as possible or no maintenance	D
Life span	10 years	D
Quality and Reliability	Five years: the Mean Time Before Failure (MTBF) and Mean Time Before Repair (MTBR) of the product	D
Manufacture	Manufacturing run between 7,500 to 20,000 units	D
Manufacturing cost	Between £3.20-£5	D
Marketing price	£16	D
Distribution	Mainstream through pharmacy chains	D
Disposal and Sustainability	Minimal number of parts	D
	Easy disassembly	D
	Contain no toxic elements	D
	Outer casing material suitable for recycling	W
Standards and Specifications	Conform to BS Kite Mark and CE legislation	D
Legal requirements	A disclaimer should be added to the user manual and packaging to avoid misuse	D

* Demand (D), Wish (W)

9.2.2 Conceptual Design

Ten force amplifying concepts were generated from a brainstorming exercise:

1. single class lever scissor style handheld solution, two moving levers;
2. single class lever arrangement, one moving lever;
3. class two lever arrangement, as in a nutcracker, two moving levers;
4. class two lever arrangement, as in a hole punch, one moving lever;
5. gear train handheld solution;
6. rack and pinion coupled with concept one;
7. vacuum to reduce pressure on foil side of blister pack;
8. air pressure pump and valve;
9. primary and secondary fluid pistons;
10. redesign of blister pack.

These concepts were evaluated using a set of criteria (market attractiveness, portability/size, mechanical advantage effectiveness, low production cost per unit, ease of use, intuitiveness, incorporation into existing infrastructure), using a rating scale. Concept 4 was selected as it met all the criteria and adopted a simple mechanical advantage movement. Consequently three more parallel products were analysed to understand the varying ways mechanical advantage could be derived from mechanisms and applied to a task. These analyses informed the embodiment design.

9.2.3 Embodiment Design

At the embodiment design stage, six card mechanisms were mocked up (Table 9. 3) to evaluate the various ways in which a hole punch style mechanism could be applied to this project.

Table 9.3. Card mechanisms

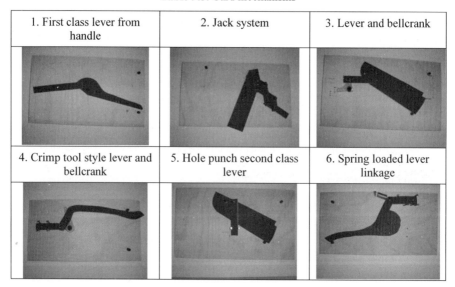

1. First class lever from handle	2. Jack system	3. Lever and bellcrank
4. Crimp tool style lever and bellcrank	5. Hole punch second class lever	6. Spring loaded lever linkage

Similarly to the concept selection, the mechanisms were evaluated using a set of criteria and Mechanism 5 was selected for further development because of:

- Its simplicity, and therefore probable low part count and easier assembly at manufacture.
- Its shorter overall length, making the final product the smallest it can be.
- Its efficient force transfer. User force is transferred directly down through the punch pin with no linkages or direction changes potentially losing force or adding weakness.
- Its intuitiveness. The affordance of the hole punch mechanism can be exploited in the final product so that the user intuitively knows how the mechanism will dispense the pill.

The weakness of the chosen mechanism was that the punch pin was surrounded by casing, significantly prohibiting visibility of this key moving part. Since a symptom of MS is reduced vision and blurring, it was paramount that the punch pin be moved to a more visible location, so that the blister pack could be placed accurately beneath it. Four proving principle prototypes (PPP's) were made to find the optimal solution. Two are illustrated in Figure 9.2.

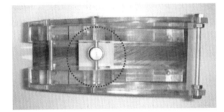

PPP1: Punch pin in the middle   PPP4: Punch pin to the side

Figure 9.2. Proving principle prototypes

Additional exploration at the embodiment stage included development of the handle, the punch head, and the tray/punch hole (Figure 9.3). To incorporate a range of pill punch heads into the product, a rotating head which can dispense large and small tablets and capsules was developed (see details in Figure 9.6).

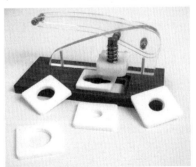

Exploration of different punch head sizes   Exploration of different tray shapes

Figure 9.3. Development of components

Users were involved in testing the PPP's (Figure 9.4) and they helped the designer to make important decisions. For example, since none of the MS users attempted to use the prototype in their hands or would want to dispense the tablets in their hands, the designer decided that the handle would not be gripped as previously intended. Instead, the product should be placed on a table surface, and force applied flat to the handle. Decisions like this were used to revise the preliminary specification to embark on the final stages of detail design.

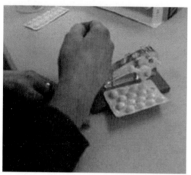 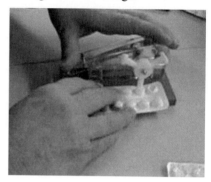

Figure 9.4. User testing

9.2.4 Detail Design

The detail design aimed to specify exact values and properties to the design and details of manufacturing. In addition, product branding and styling were considered. The final dimensions of the artefact were 137mm(L)× 80mm(H)×77mm (W) (Figure 9.5), which proved comfortable for users with MS to use as a stationary pill dispenser.

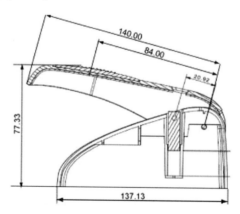

Figure 9.5. Final dimensions

Design for manufacture was carefully considered on the basis of the revised product design specification and full details can be found from (Vanns, 2007). Since the product was planned to be sold through a mainstream retailer, the most prominent pharmacy brands on the high street in Britain were studied: Superdrug and Boots. Boots has been established far longer than Superdrug, and has as a result gained the trust of consumers, particularly the older people. It was felt that branding this product through Boots would be most suitable. After a detailed brand analysis, it was decided that the Boots 'Healthclub' was the best sub brand to use for the product, as it is linked to the same area of the store in which medication is sold. The product logo and colours were then taken from this Boots 'Healthclub' brand. Mood boards illustrating 'independence', 'accomplishment' and 'self-worth' and anti mood boards illustrating 'frailty', 'dependence' and 'helplessness' were made to give direction to styling. Consequently product details were defined, such as the small raised domes on the handle to indicate the intended area of force application to the user (Figure 9.6)

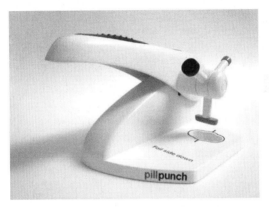

Figure 9.6. Aesthetic model of the 'pillpunch'

9.3 User Testing and Evaluation

The user testing of the fully functional prototype showed that the pill punch was successful in aiding MS sufferers to dispense medication in a hassle and pain free way. As an MS user commented "That was easy, very easy, I didn't need much pressure for that at all". The pill dispenser accommodates various of ways of applying force, thus has the potential to benefit a wider range of users in addition to MS sufferers.

The working prototype was also tested by non MS users. One of such participants attempted to dispense a pill from the blister pack the wrong way up: foil side up. Therefore the punch head would simply make contact with the foil and push it into the back of the blister pack. To overcome this error, a 'foil side down' note was added to the surface where the blister pack was placed.

In addition to user testing, the product had been evaluated through concept selection and prototype iterations, so that the ease of use, aesthetics and desirability had been fully explored through the design process. The user feedback further verified the product's acceptability. The final product has met all the specifications set at the PDS. By amplifying user force input by a factor of three, it significantly reduces the force requirement from the user, thus removing stress, frustration and painful grip procedures from the dispensing process. None of the competitor products on the market has this advantage.

9.4 Discussion and Conclusions

Despite the packaging standards in different European countries (such as BS 8404, 2001 and DIN 55 559, 1998) addressing accessibility issues and recent research on medical packaging (such as those published in the European Design for All e-Accessibility Network www.education.edean.org/pdf/Case032.pdf and the Helen Hamlyn Centre www.hhrc.rca.ac.uk/archive /hhrc/programmes/ra /2005/p9 .html), blister packs of medication still present an accessibility challenge to many patients with weak hands or dexterity problems. By providing a 'universal' pill dispensing solution suitable for all types of tablets and capsules, the pillpunch has helped solve the problem of not only people with MS, but many others who are challenged by blister packs.

This project reflects the underlying philosophy of inclusive design, *i.e.* by designing for weaker users, you include more potential users, as put by the late Bernard Isaacs, Founding Director of the Birmingham Centre for Applied Gerontology "design for the old and you include the young". By targeting female MS sufferers who have less strength and smaller hands and applying mainstream aesthetics and design principles, the designer has created a socially inclusive, technically sound, and commercially viable pill dispenser that appeals to all: the product has received favourable comments from the general public at the Graduation Show 'MADE IN BRUNEL' and national design show "New Designers". General conclusions regarding inclusive design can be drawn from this project:

- inclusive design starts with a deep understanding of the users targeted;
- users should be involved throughout the design process, not only for testing, but also for their subjective preferences;
- the inclusive design process can follow a standard systematic engineering design process, but designers should be open to changes when their intention proves contradictory to users' preferences;
- inclusivity is achieved by removing/reducing task requirements to accommodate user capabilities, whilst maintaining mainstream aesthetics.

9.5 References

Clarkson P, Keates S (2003) Quantifying design exclusion. In: Keates S, Langdon P, Clarkson PJ, Robinson P (eds.) Universal access and assistive technology. Springer, London, UK

Pahl G, Beitz W (1996) Engineering design – a systematic approach, 2nd edn., translated by Wallace K. Springer, London, UK

Pugh S (1991) Total design: integrated methods for successful product engineering. Addition-Wesley Publishing Company, Wokingham, UK

Vanns N (2007) Pillpuch, major report. School of Engineering and Design, Brunel University, UK

Chapter 10

Prior Experience of Domestic Microwave Cooker Interfaces: A User Study

T. Lewis, P.M. Langdon and P.J. Clarkson

10.1 Introduction

Previous work has considered how the performance levels for daily living product such as a motor car and a digital camera vary with age, generation, cognitive ability and previous experience (Langdon *et al.*, 2007). For both, it was found that there was clear evidence to indicate a reduction in performance with increasing age and an improvement in performance with increasing cognitive ability. The analysis and classification of types of errors made was dominated by response selection; the selection of the wrong response as an action given a particular state of the interface. However, overall, the strongest correlations were found to be between performance measures such as times to complete tasks and the extent of previous experience. Building on the findings of the previous study, 16 users were recruited to complete a set of trials with two microwave cookers from the same manufacturer; one with a dials interface and the other with a buttons interface.

10.1.1 Inclusive Design

It is predicted that by 2020 more than 50% of the UK population will be over the age of 50 and the over-80 age range will be growing the most rapidly (Keates and Clarkson, 2004). To comply with the ideals of inclusive design, a designer must ensure a product minimises the number of people excluded or who experience difficulty with the product (Nicolle and Abascal, 2001). Inclusive products can reduce the problems suffered by older people and those with impairments but frequently they are also more useable by everyone. Guidance for products for use with people with specific disabilities exists from similar research areas (TRACE Center, 1992; Poulson *et al.*, 1996) but designers prefer more concise, general advice (Dong *et al.*, 2004).

10.1.2 Memory and Learning

Prior experience is determined by a user's ability to acquire, store and retrieve relevant information in their long term memory (LTM). A human information processing approach to cognition can model these processes and considers carefully the roles played by both working memory (WM) and LTM (*e.g.* Baddeley, 2000). These models also assume an executive function as part of the WM that is able to manage the interface between the processes. According to generally accepted theory and experimental evidence, (Baddeley, 2000) items stored in the WM can persist for around 10-15 seconds. The WM has been shown to hold sufficient, separate items for further consideration from either perception, memory or other input sources. It is hypothesised to consist of four components, all of which are likely to be extensively involved in normal product interaction

- a central executive to divide attention amongst the required tasks;
- a phonological loop for speech based information;
- a visuospatial sketch pad for organised, visual information;
- an episodic buffer, that binds information into unitary episodic memories.

10.1.3 Previous Experience and Training Transfer

New products are often an evolution of a previous design or make strong reference to products that have gone before them. It is assumed that the more experience a user has of similar products, the quicker they will learn the operation of a new one.

Research into training transfer has considered the effect of similarity, between the training situation and the one being tested, on retention of training. This can become counter-productive when the training is too similar to the actual new product and acquired accepted behaviour on the training product now represents an error on the actual product. Training with a flight simulator will offer improved performance over watching a video of a pilot operating the flight deck (Lintern *et al.*, 1990). Simplifying an interface in the early stages of learning makes it easier to use (Carroll and Carrithers, 1984). In the case of consumer products, a designer may alter the position of particular controls causing proactive interference for users familiar with an older design.

10.1.4 Generational Effects

A previous study (Lewis and Clarkson, 2005) examined how well certain standard symbols across product families were recognised by different generations. It was found that for older generations some modern symbols would go completely unnoticed. Studies in The Netherlands have explored this further by outlining technological periods in the past 100 years that categorise generations by the types of interface they were familiar with when young. In particular, layered computer interfaces were shown to be most suited to the learning processes of those 25 and younger (DoCampo, 2000). The later generations' familiarity with layered displays and menus is particularly relevant to some consumer products of today that fail to include older users.

10.2 Background

A previous study, (Langdon *et al.*, 2007), trialled a number of users on a motor car and a digital camera. Users were given a task list for each product and asked to complete as many tasks as possible. Cognitive ability was assessed by a combined cognitive scores test and previous experience was measured by means of a usage questionnaire. Age, generation, cognitive ability and previous experience all correlated with performance, with increasing previous experience and decreasing task times giving the strongest correlations. The correlation for the car was notably high and one conclusion of this study suggested that across the different car manufacturers there were more common standards, consistent symbols and interface controls than in the camera market. A generational effect was anticipated in the over 70's similar to that found by Docampo (2000) such that display and menus would suit generations used to newer technology in performing with the camera. This was anticipated as a step change in performance. The over-70's however were not able to complete many of these tasks and the generational effect manifested itself by failure to complete the tasks. To encourage more users to make attempts and to avoid open-ended task segments, a timing cap was introduced. Errors made by the users were also examined. In the case of the camera, errors correlated well with times yet with the car high times often corresponded with few errors. Users seemed to consider the options available to them in the interface for longer with the car before resorting to a trial and error approach. Most errors made across both products were identified as "random selection" in that when the user was unsure how to complete a task they attempted many different alternative actions, looking for any interface response.

Whilst this study compared performance levels on two quite different products, a follow-on study could helpfully consider one type of product and the effects of varying only the interface controls. A microwave was chosen as the product with the interface variations of a dial controlled model and a button controlled model. It was predicted that all age ranges would perform better with the dials interface. However, the difference in performance between the two models should increase with age as a result of a generational effect in the older users.

10.3 Method

10.3.1 Users

Sixteen participants aged from 21 to 85 years old were selected to participate in the microwave trials. Eight were male and eight were female; three additional participants were used for control trials. All were informed the trials were being conducted to examine the products' performance with a range of users rather than to compare individual users' performances. They were informed they could stop the trial at any point and all their recordings could be deleted on demand. The detailed methodology followed the codes of practice of the British Psychological Society (BPS Ethics, 2006).

10.3.2 Cognitive Assessment

Each product user attempted a 15-minute combined cognitive test based on multi-scale intelligence testing (CCS20, 2006). Users were asked to attempt as many questions as possible but could pass or guess at questions when they were unsure of the correct answer. The test provided a combined cognitive capability score as well as scoring for the individual sub-scales; verbal and mathematical ability, spatial, logic and pattern recognition, general knowledge, STM, and visualisation and classification. For simplicity, these were combined to form four sub-scales; perception, reasoning, STM and LTM. The overall score contained an age-correction factor. However, as in this case trial performance was expected to vary with age, the effect of this factor was removed, scoring all users as if they were aged 20 and the test renamed to combined cognitive score 20 (CCS20).

10.3.3 Experience Testing

In the previous study, the experience of a user had been measured by a questionnaire. Questions were weighted differently to account for their perceived usefulness in helping the users with the new product. For the current experiment, however, experience was tested on a knowledge basis. Hence, users answered questions on symbol recognition and button position. These included some symbols that would be found on the microwaves that were tested, as well as those from other microwaves. This was intended to provide a more objective measure of a person's experience levels. Users also retrospectively completed a usage experience questionnaire, as before, so both sets of measures could be compared. (See Figure 10.1).

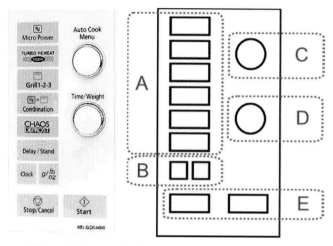

Figure 10.1. Sample from the microwave experience test

10.3.4 Products

Two branded microwaves from the same manufacturer were used for the trials. In their outward appearance, the only difference between the two microwaves was the user interface control panel. The ME20S interface consisted of ten numbered buttons, three function buttons and two further buttons representing 'activate' and 'cancel'. The M20S model's interface had two dials; one for a power setting and the other for timing (see Figure 10.2). As part of a range offered by a UK electrical retailer, they were both within the lower priced microwave range, with the buttons model being slightly more expensive.

Figure 10.2. The microwave dials and buttons interfaces

10.3.5 Segmentation and Errors

Users were provided with a list of tasks to be attempted for the trial that was identical for each microwave. Each task was provided on a separate page of a booklet which gave clear a indication to the observer that they had finished attempting the current task and were progressing to the next. The trials were video recorded and the users invited to watch their performance afterwards. During this time they were asked to provide a retrospective audio commentary on the difficulties they felt they had encountered and explain why they had made specific actions. Any actions that deviated from a standard procedure were deemed an error. Where there was more than one sequence of events that achieved the correct result, errors were only recorded if the action brought the microwave further from the required state or were unnecessary. An errors analysis was performed to check whether users were trading off speed against accuracy.

10.3.6 Trials

The written components of the trials were completed first to avoid any knowledge gained during the trial affecting the users' performance in the experience testing. The order in which the two microwaves were tested alternated to avoid any ordering effect. One task required the same procedure of actions for both microwaves (open and close the door) as the handle used to achieve this was identical and appeared in the same position on both models. Participants were limited to a maximum of 5 minutes to complete each task. The task list was:

1.   open and close the door;
2.   set the power to 50%;
3.   set the timer to 30 seconds and activate;
4.   set to defrost mode;
5.   set the time to 1 minute, activate and then stop with 50 seconds remaining.

10.3.7 Complexity Control

The buttons model had 15 controls compared to the two controls on the dials model. To ensure that any differences in performance between the two models were not simply a result of this additional visual, motor and cognitive complexity, three additional users were recruited to take part in control experiments. The buttons model was modified, covering all unnecessary buttons for the tasks with the exception of "cancel". This was the "buttons restricted" condition.

10.4 Results and Analysis

10.4.1 Ageing

The strongest result was the anticipated aging effect on times to complete tasks and the difference in performance level between the two microwaves (see Figure 10.3). The dials model allowed all users to perform the tasks quicker than the buttons model. A paired scores t-test confirmed this (t = -5.47, df = 15 p < 0.001, 2-tailed). The small number of the control group precluded statistical analysis but a regression line suggested an intermediary relationship. The difference between the two microwaves is relatively small for the youngest users but grows steadily as age increases (dials mean = 204 sec, buttons mean = 456 sec). There is some evidence for a generational effect with the eldest users over 70 taking substantially longer than the trend suggests. The dials model has a little variation from this trend with users' performances being quite predictable, as the distribution is less variable. The buttons model has far more variation reflecting the trial and error loops that users entered into as they attempted to complete the tasks. The variation increases with age. A linear model accounted for significant amounts of the variance of time with age (Dials: R2 = 0.585, F =19.7, df = 15, p<0.001. Buttons: R2 = 0.717, F = 38.5 df = 15, p<0.001). A separate analysis was carried out for errors, which, as in previous experiments, proved to be similarly distributed to the timing data with respect to age. There were significant differences between dials and buttons with their effects on errors and this was confirmed by both a paired-score and independent groups t-tests (t = -4.51, df = 15 p < 0.001, 2-tailed). There were a great deal more errors in the buttons condition (dials mean = 1.5, buttons mean = 40), reflecting the greater number of incorrect actions made with the complex buttons interface.

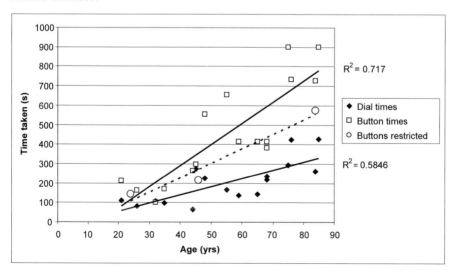

Figure 10.3. Time taken performance with age

10.4.2 Experience Analysis

There was far less correlation between the experience score generated on the basis of knowledge and time performance (see Figure 10.4). A linear model failed to account for significant amounts of the variance of time with experience (Dials: R^2 = 0.289, F =5.7, df = 15, n.s. Buttons: R^2 = 0.202, F = 3.5 df = 15, n.s). The knowledge based experience test produced a better correlation for the dials model (R = 0.54) than the buttons one (R = 0.45) but they are both weaker than that attained previously for the cars and cameras equivalent experience graphs. The cars and cameras experience scores had been taken from an experience usage questionnaire rather than the knowledge test. An even weaker result was obtained for Usage experience (see Figure 10.5).

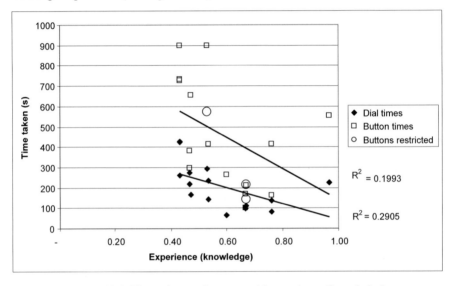

Figure 10.4. Time taken performance with experience (knowledge)

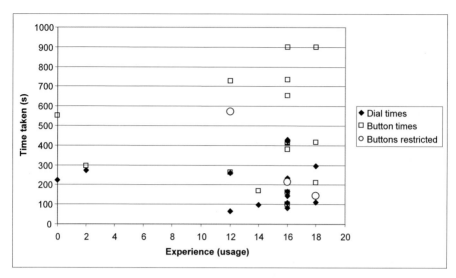

Figure 10.5. Time taken performance with experience (usage)

10.4.3 Cognitive Analysis

The CCS20 with performance correlates well and is comparable to those from the cars and cameras trials (see Figure 10.6). As predicted, all users perform better with the dials than the buttons across the cognitive scale. (Dials: $R^2 = 0.330$, F $=6.9$, df $= 15$, $p<0.05$. Buttons: $R^2 = 0.465$, F $= 12.14$ df $= 15$, $p<0.05$). The difference between the performances narrows with increasing cognitive capability. Results for the breakdown of the individual cognitive components show similar correlations for both the perception and reasoning scales. STM is particularly strongly correlated with performance whereas LTM is particularly weak. This is again consistent with the cars and cameras experiment.

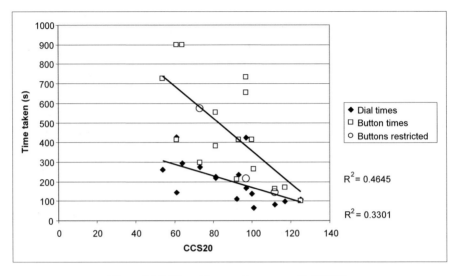

Figure 10.6. Time taken performance with CCS20

10.5 Discussion

The results are consistent with the predictions from the hypotheses. There were significant differences between times to complete tasks for the dials and buttons interfaces, and these were not due to a time-accuracy trade-off. A significantly larger number of errors were also recorded for the buttons interface. It is presumed that this is due to their being a greater number of alternative control actions with the buttons, combined with reduced feedback of results of actions. Aging shows the strongest correlation with times, and all users perform better with the dials model than the buttons version. Younger users were proficient in using both microwaves and quickly learnt the sequencing necessary in the buttons model. The differences in times between the two interfaces were greater with increasing age. The buttons version gave rise to much more variation in the times, particularly as age increased. This appears to reflect the adoption of a trial and error approach with users spending more time trying alternative options that were incorrect. The three users who trialled the restricted buttons control task performed better than similarly aged users on the full buttons model but their times to complete the task were not as low as for the dials microwave. This suggests that the additional complexity of the buttons interface was not in itself the cause of the longer times or greater errors.

There was evidence of a generational effect with some participants unable to complete certain tasks with the dials microwave. Prior experience measure showed little correlation for either the knowledge or the usage variant, against task performance. This weak correlation suggests that the buttons microwave interface may require very specific knowledge in order to use it and this may be exacerbated by the lack of standards and conventions across the product market. No users had used a product by this manufacturer before and this may have affected the results of the usage questionnaire.

There was a strong relationship between the cognitive measures and performance for both interfaces but particularly for the buttons. This suggests that increasing cognitive capability enabled increasingly effective use of the controls and that greater variability occurred with the more complex alternatives and response selection with the buttons interface. A high STM correlation also indicated that working memory was significant, particularly with the buttons model where several actions had to be performed in a sequence. Errors were found to have similar distributions to times taken, ruling out a speed-accuracy trade-off, but in all cases time as a performance measure produced a stronger correlation. Errors were significantly higher with the buttons model than the dials. Of the errors that did occur with the dials interface, many were recorded as perceptual errors caused by not seeing the dials clearly and selecting, for example, 30 minutes rather than 30 seconds.

10.6 Conclusions

To summarise the findings, dials were generally found to lead to better performance with all users. The differences between dials and buttons in performance were such that dials were found to be clearly easier to use for both younger users and also those with higher cognitive ability. A control group for complexity of the interface produced the expected intermediary result suggesting that the larger number of control buttons directly affected the difficulty of use. Cognitive capability was closely related to ability to use the interface and the effect of previous experience was found to be less important for microwave controls than in the previous cars and cameras experiment. The dials model outperformed the buttons model in terms of times to complete actions and number of errors with all users. It is interesting to note that the retailer at which the microwaves were purchased offered nineteen different button based models yet only two dials models. The dials model is comparable to the car interface used in the previous experiment (Langdon et al., 2007) in that high consistency in the interface and simplicity of the design throughout the product market meant that the users were able to complete the trials faster and with fewer errors. The buttons microwave model was more similar in effect to a digital camera interface where there were many different symbols and control layouts that could reduce even experienced users to a low level of performance (Langdon et al. (2007). However, ongoing work is now focussing on the differences between the features of the different interfaces that leads to greater difficulty of learning and use.

Aging was a more significant factor here whereas experience had been more dominant in the previous study. This may be partly due to greater familiarity with dials interfaces for the older users but could also reflect their cognitive difficulty with the buttons interface with its added perceptual and cognitive demand. This suggests that it is likely that some classes of product exist for which prior experience of other products with similar features, or from within the same market, will be more effective, useful and relevant to their usability than for other classes. In looking further for explanations for these findings it will clearly be desirable to quantify the extent to which variability in the timing and error results in both experiments were due to contributions from ageing, experience and cognitive capability factors.

10.7 References

Baddeley AD (2000) The episodic buffer: a new component of working memory? Trends in Cognitive Sciences, 4(11): 417–423

BPS Ethics (2006) Available at: www.bps.org.uk/the-society/ethics-rules-charter-code-of-conduct/code-of-conduct/code-of-conduct_home.cfm (Accessed in December 2007)

Carroll JM, Carrithers C (1984) Blocking learner error states in a training wheels system. Human Factors, 26(4): 377–389

CCS20 (2006) Available at: www.intelligencetest.com (Accessed in July 2006)

Docampo RM (2000) Technology generations handling complex user interfaces. PhD-thesis, TU Eindhoven, The Netherlands

Dong H, Keates S, Clarkson PJ (2004) Industry perceptions to inclusive design – a comparative study. In: Proceedings of the 2004 ASME International Design Engineering Technical Conferences (DETC'04), Salt Lake City, UT, US

Keates S, Clarkson PJ (2004) Countering design exclusion – an introduction to inclusive design. Springer, London, UK

Langdon PM, Lewis T, Clarkson PJ (2007) The effect of prior experience on the use of consumer products. Universal Access in the Information Society, 6(2): 117–217

Lintern G, Roscoe SN, Sivier J (1990) Display principles, control dynamics, and environmental factors in pilot performance and transfer of training. Human Factors, 32: 299–317

Lewis T, Clarkson PJ (2005) A user study into customising for inclusive design. In: Proceedings of Include 2005, Helen Hamlyn Research Centre, London, UK

Nicolle C, Abascal J (eds.) (2001) Inclusive design guidelines for HCI. Taylor and Francis, London, UK

Poulson D, Ashby M, Richardson SJ (1996) USERfit: a practical handbook on user centred design for assistive technology. HUSAT Research Institute for the European Commission

TRACE Center (1992) Accessible design of consumer products. Available at: http://trace.wisc.edu (Accessed in December 2007)

Chapter 11

Prior Experience and Intuitive Use: Image Schemas in User Centred Design

J. Hurtienne, K. Weber and L. Blessing

11.1 Inclusive Design and Intuitive Use

Inclusive design is concerned with the "design of mainstream products and/or services that are accessible to, and usable by, people with the widest range of abilities within the widest range of situations without the need for special adaptation or design" (BSI, 2005). What does "the widest range of abilities" of people refer to? Usually, users' abilities are categorized into sensory, motor, and cognitive abilities (Keates and Clarkson, 2003).

Supporting sensory abilities means enhancing the perceptibility of user interface elements (*e.g.* enhancing contrast and font size). Supporting motor abilities means enhancing operability of controls (*e.g.* changing their size and spacing or the force needed to operate them). Supporting cognitive abilities means enhancing thinking and communication (Clarkson *et al.*, 2007) and is often referred to as enhancing 'intuitive use' (*e.g.* Story, 1998). However, it often is not clear what intuitive use is and how designers can design for it (Blackler, 2006).

Intuitive use is defined as the *users' subconscious application of prior knowledge that leads to effective interaction* with a product (Hurtienne and Blessing, 2007; Naumann *et al.*, 2007). When knowledge is so deeply rooted in memory that it is applied automatically and subconsciously, it means that interaction takes place with a minimum of mental effort.

What prior knowledge on the part of users can designers of interactive systems focus on? Different sources of knowledge are possible. These sources can be roughly classified along a continuum. The first and lowest level of the continuum consists of *innate* knowledge that is 'acquired' through the activation of genes or during the prenatal stage of development. Generally, this is what reflexes or instinctive behaviour draw upon. Purists will see this as the only valid level of knowledge when talking about intuitive interaction, because it assures universal applicability and subconscious processing.

The *sensorimotor* level consists of general knowledge, which is acquired very early in childhood and is from then on used continuously through interaction with the world. Children learn for example to differentiate faces; they learn about gravitation; they build up concepts of speed and animation. Scientific notions like affordances (Gibson, 1979), gestalt laws (*e.g.* Koffka, 1935), and image schemas (Johnson, 1987) – which are the topic of this paper – reside at this level of knowledge.

The next level is about knowledge specific to the *culture* in which an individual lives. This knowledge can vary considerably between cultures and may influence how people approach technology. It touches, for instance, the realm of values (*e.g.* what constitutes a taboo), the styles of visual communication (*cf.* Japanese manga vs. American comics), but also concerns knowledge about daily matters like the usual means of transportation (*e.g.* busses, trains, or bicycles) or the prevalent form of energy supply (*e.g.* by a public power line or by burning wood for heating).

The most specific level of knowledge is *expertise*, *i.e.* specialist knowledge acquired in one's profession, for example as a mechanical engineer, an air traffic controller, or a physician; and in hobbies (*e.g.* modelling, online-gaming, or serving as a fire-fighter).

Across the sensorimotor, culture, and expertise levels of knowledge, knowledge about *tools* can be distinguished (Hurtienne and Blessing, 2007). Tool knowledge is an important reference when designing user interface metaphors. The desktop metaphor, for instance, is thought to tap the knowledge of a typical office environment including folders, documents, and a wastebasket.

The application of knowledge may be subconscious from the beginning on (as with reflexes) or may have become subconscious due to very frequent exposure and reaction to stimuli in the environment: the more frequent the encoding and retrieval was in the past, the more likely it is that memorised knowledge is applied automatically and subconsciously. Knowledge at the expertise level is acquired relatively late in life and is (over the life span) not as frequently used as knowledge from the culture or sensorimotor level. Knowledge from the lower levels of the continuum is therefore more likely than knowledge from the upper levels to be applied automatically. If the subconscious application of knowledge is a precondition for intuitive use, it will be more likely to see intuitive interaction involving knowledge at the lower levels of the knowledge continuum. Also, the further down in the continuum, the larger and more heterogeneous the user groups that can be reached. Designs aiming at the lower levels of the continuum therefore will be more inclusive.

In the remainder of this article image schemas, rooted in the sensorimotor level of knowledge, are presented as a framework for user-interface design. Some early applications of image schemas are discussed.

11.2 Image Schema Theory

According to Johnson, one 'father' of image schema theory, image schemas are abstract representations of recurring dynamic patterns of bodily interactions that structure the way we understand the world (Johnson, 1987). The UP-DOWN image schema, for example, forms the basis of "thousands of perceptions and activities we experience every day, such as perceiving a tree, our felt sense of standing upright, the activity of climbing stairs, forming a mental image of a flagpole, measuring the children's heights, and experiencing the level of water rising in the bathtub" (Johnson, 1987). The UP-DOWN image schema is the abstract structure of all these experiences. It is neither a concrete mental image nor a meaningless abstract symbol. Depending on the author, about 30 to 40 of such image schemas are distinguished. Table 11.1 show the image schemas that were used in the studies referred to and described in this paper.

Table 11.1. List of image schemas used in the studies referred to and described in this paper

Group	Image Schemas
BASIC	OBJECT, SUBSTANCE
SPACE	CENTER-PERIPHERY, CONTACT, FRONT-BACK, LEFT-RIGHT, NEAR-FAR, PATH, ROTATION, SCALE, UP-DOWN
CONTAINMENT	CONTAINER, CONTENT, FULL-EMPTY, IN-OUT, SURFACE
MULTIPLICITY	COLLECTION, COUNT-MASS, LINKAGE, MATCHING, MERGING, PART-WHOLE, SPLITTING
PROCESS	CYCLE, ITERATION
FORCE	ATTRACTION, BALANCE, BLOCKAGE, COMPULSION, COUNTERFORCE, DIVERSION, ENABLEMENT, MOMENTUM, RESISTANCE, RESTRAINT REMOVAL
ATTRIBUTE	BIG-SMALL, DARK-BRIGHT, HEAVY-LIGHT, STRAIGHT, STRONG-WEAK, WARM-COLD

Not only SPACE image schemas like UP-DOWN are found in this table, but also image schemas for CONTAINMENT, MULTIPLICITY, PROCESS and FORCE. The BASIC image schemas OBJECT and SUBSTANCE underlie much of human abstract reasoning. For example, ideas are conceptualized as concrete OBJECTS expressed in "I can't *grasp* the idea" or "Sally *carries* that idea *around* with her all the time". ATTRIBUTE image schemas are less rich in structure and denote common properties of objects.

This transfer of image schemas from physical interaction with the world to the thinking about abstract, non-physical entities is called *metaphorical extension* of the image schema. Metaphorical extensions are often grounded in bodily experience that is universally accessible to all people. For example, experiencing the level of liquid rising in a container when more liquid is added or seeing a pile of paper shrink when sheets are taken away leads to the metaphorical extension

MORE IS UP, LESS IS DOWN of the image schema UP-DOWN. This correlation of amount and verticality subsequently is generalized to non-physical abstract entities like money or age, for example in expressions like "My income *rose* last year", "Rents are going *up*", or "He is *under*age". Other metaphorical extensions of the UP-DOWN image schema are GOOD IS UP, BAD IS DOWN ("We hit a *peak* last year, but it's been *down*hill ever since"), HAPPY IS UP, SAD IS DOWN ("I'm feeling *up*", "He is really *down* these days"), or HIGH STATUS IS UP, LOW STATUS IS DOWN ("She'll *rise* to the top", "He's at the *bottom* of the social hierarchy"). These and other metaphorical extensions of image schemas have also been validated outside linguistics, for example in gesture research, psychology, and computational neuroscience (for an overview see Hampe, 2005), and also in user-interface design (Hurtienne and Blessing, 2007).

11.3 Image Schemas in User-Centred Design

Image schemas contribute to intuitive interaction via the principles of spatial and abstract mappings. *Spatial mappings* can occur between user interface controls and expected effects in the real world or between displays and controls. Turning the steering wheel of a car to the LEFT will result in a leftwards motion, turning it to the RIGHT results in a rightward motion of the car. *Abstract mappings* occur between user interface elements and abstract concepts via metaphorical extensions. Examples are using UP-DOWN in a vertical slider for controlling the intensity of the speaker volume (MORE IS UP) or rating the attractiveness of a new car (GOOD IS UP). This use of image schemas for representing abstract concepts is one of the major promises for user-interface design, because, in the mind of users, they subconsciously tie the location, movement and appearance of UI elements to their functionality.

How can a designer use image schemas when designing intuitive interaction? Two ways are possible: (1) Designers can apply image schemas during the analysis and design phases of a user-centred design process. A proof-of-concept case study has been conducted to show the applicability and usefulness of image schemas for user-interface design. (2) Designers can exploit image schema uses that are already documented. A database of image schemas and metaphorical mappings has been built as a first step to support this. Both approaches are discussed in the following sections.

11.3.1 Image Schemas in a User-centred Design Process

The user-centred design process, according to ISO 13407, comprises four core activities: (1) Understand and specify the context of use, (2) Specify the user and organisational requirements, (3) Produce design solutions and (4) Evaluate designs against requirements. A proof-of-concept study was set up that undertook the redesign of the invoice verification and posting procedure in the accounting department of a German beverage company. The design process, the methodology and the outcome of the study are discussed within this section.

Activity 1: Understanding and Specifying the Context of Use

In this first activity, and according to the ISO standard, each of the following were analysed: users, their tasks, the existing technology, and the socio-technical context in which they interact.

Three users were interviewed and observed at their workplace. During work analysis users were asked to think aloud and their utterances were recorded using a voice recorder. The users were between 42 and 55 years old and had between 2 and 17 years of experience with the task. The workplace of each user has two screens. The left screen contains an SAP R/3 application displaying invoice data. The right screen holds a digitised image of the original paper invoice, a list of contact persons and a tool for writing and forwarding notes.

The task of invoice verification and posting is not overly complex. The enterprise purchases goods or services and receives an invoice from the supplier in return. The user's task starts with comparing the data in the SAP R/3 application with the original invoice image and making sure all information is complete and according to legal requirements. After that prices and quantities are checked against orders and receipts of goods. If everything is correct, the users are enabled to post the invoice. If information is missing, inappropriate or unclear, users forward the document to the appropriate operating department. There the data or authorisations they need to finally post the invoice are provided.

For the specification of the context of use image schemas were extracted from (1) single work steps, (2) the user interface of the current system, (3) the user's interaction with the system, and (4) the user's mental model. The latter was achieved by transcribing and analysing the language of the users talking about their work. Image schemas could indeed be used to describe most aspects of the context of use. For example, one of the first steps the users undertake is to check whether the data of the paper invoice was digitised correctly. In order to do so, users compare the digitised data with the original invoice image. This step is linked to the image schema MATCHING because the two different sets of data have to fit together – they have to match. The user's language supports this image schema (German): "Dann muss man halt alles *abgleichen*, [...] ob das alles *stimmt*". ("Then one just has to *compare* everything, [...] to see if everything *matches*").

Activity 2: Requirements Specification

Requirements were specified with a focus on (1) requirements resulting from business and financial objectives and (2) user requirements such as the allocation of users' tasks, user-interface design, communication and co-operation between users. The requirements were directly derived from the results of the context-of-use analysis. The basis for formulating the requirements was the work steps that were indispensable to the invoice verification and posting process plus the related image schemas. In a second step, image schemas that occurred exclusively in the user's mental model were added.

Three sets of requirements were created, each focussing on a coherent set of sub-tasks of the user. This included requirements for (1) opening invoices from the in-box, (2) verifying and posting of invoices and (3) forwarding of invoices. The earlier presented example of the extracted image schema MATCHING resulted (among others) in the following requirements: "Display digitised data and original image of invoice at the same time." and "Support the user in comparing and matching digitised data against the original image of the invoice." These requirements were linked with the image schema MATCHING (in addition, the image schemas SPLITTING, CONTAINER, CONTENT and FULL-EMPTY were extracted from this work step). Note that although this requirement seems rather trivial, the current system does not provide any solution for it. The users spend unnecessary time and energy with the often hard-to-read image of the original invoice to find the data – such as bank details – they have to compare.

Activity 3: Producing Design Solutions

The third phase is highly iterative as it is tightly coupled with the next activity, evaluation. Design solutions start with rough sketches of the task flow in the system and end with the finer levels of detail in the user interface. The redesigned solution consisted of three main screens that were in line with the three sets of requirements above.

We assumed that image schemas can work as a helpful tool to "translate" requirements into concrete design solutions. Before synthesising single design solutions into the overall user interface of the system, we tried to focus on every requirement independently. For this we used the creativity technique of the morphological box. For each image schema several alternative solutions were sketched. As an example Figure 11.1 shows three alternatives that instantiate the earlier introduced MATCHING-requirement.

After this step the best fitting individual solutions for every requirement were selected and put together into the three screens that were designed to improve the user's current system. The MATCHING-requirement was integrated following alternative two in Figure 11.1. When users want to find corresponding data on the original invoice, they simply put the mouse pointer over the data fields concerned. A coloured highlight that is present in the data screen and the matching area on the invoice image is cutting down search time. This function is possible, because the system already reads the data directly from the digitised invoice. Thus the exact location of the data is known to the system.

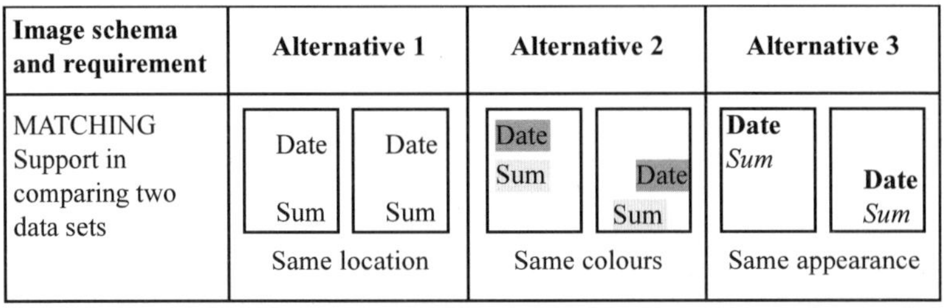

Image schema and requirement	Alternative 1	Alternative 2	Alternative 3
MATCHING Support in comparing two data sets	Date Date Sum Sum Same location	Date Sum Date Sum Same colours	Date Sum Date Sum Same appearance

Figure 11.1. Exploring design alternatives for the MATCHING image schema

Image schemas were used for realising all three sets of requirements. It became obvious during this phase how helpful image schemas are as a meta-language for translating requirements into proper design solutions. Image schemas provide abstract descriptions of what to achieve in a user-interface design. In that way they help to guide the design work without being overly restrictive.

Activity 4: Evaluation of Design Solutions
The fourth activity – evaluate the designs against requirements – is already part of the other activities of context-of-use analysis, requirements specification, and design solution sketching. In the evaluation of the existing system (Activity 1) image schemas proved to be really helpful. The evaluation also revealed many inappropriate uses of image schemas in the current system. For example, data was SPLIT across several screens and many selection and confirmation steps had to be done for no apparent reason (RESISTANCE). By simply comparing image schema instances from the system and the user-system interaction with image schemas linked to requirements, strengths and weaknesses of the current system were revealed. Missing or obsolete instances of image schemas were seen as points where further adjustment and improvement were necessary. This method also provided a first indication of how well the re-designed system fits the requirements.

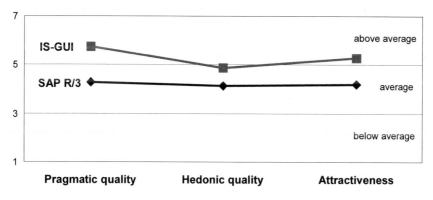

Figure 11.2. Evaluation results obtained with the AttrakDiff questionnaire (comparing the existing system with the redesigned solution)

Nevertheless, feedback from real users is crucial in judging the quality of the produced design solutions. Five users, the three users observed during context-of-use analysis and two of their colleagues, took part in the evaluation. A simplified version of the pluralistic usability walkthrough technique (Bias, 1994) was used. The users were introduced to the image schematic re-designed screens and their functions. Then they were asked to comment on the prototypes and voice their opinion. The users considered the redesigned solution more efficient, because it minimised work steps and reduced confusion. Among others, the introduced matching-function was well approved by the users and seen as a useful feature. After the walkthrough, the AttrakDiff questionnaire was administered (Hassenzahl *et al.*, 2003). It measures the pragmatic quality, the hedonic quality, and the attractiveness of an interactive product. Figure 11.2 shows the comparison between the ratings of the existing SAP R/3 system and the newly designed solution. All three criteria are improved by the image schematic redesign. The differences in pragmatic quality, hedonic quality, and attractiveness are statistically significant at an alpha level of .10 (Fisher-Pitman randomisation test, p=.016, .071, and .056, respectively; N1=4, N2=5).

11.3.2 ISCAT – Documentation of Image Schema Use

To support widespread application of image schemas in a user-interface design process, an online database of image schemas was built. (Please contact the first author to gain access). The database, called ISCAT, has three purposes. Firstly, definitions of image schemas are collected in a single place allowing user-interface designers to consult them if needed. Although the names of image-schemas seem to be self-evident we found that, when extracting them during context-of-use analysis, questions about their specific nature arose (*e.g.* what is the exact difference between a part-whole and a collection image schema?). Providing definitions in the database thus aims at reducing the subjectivity and enhancing the comparability of image schema extraction during context-of-use analysis.

Figure 11.3. Image schema database ISCAT, view of a user interface example

Secondly, designers are provided with examples of image schema applications that they may use for inspiration. For this purpose, a wide variety of user interfaces has been analysed and image schema instances have been extracted (*e.g.* of airplane cockpits, cash- and ticket machines, tangible user interfaces, business software). The database provides brief descriptions of each instance, the effect that it achieves, the strength of the effect, and whether it contributes to or hinders usability (Figure 11.3). Additionally, metaphorical extensions of image schemas as they are analysed in the cognitive linguistics literature are provided. Providing linguistic data aims at stimulating the process of finding design solutions in case no user interface examples are available in the database.

Thirdly, the accumulation of a large number of image schema instances of user interfaces can be used for further research into general rules of image schema application. Thus, implicitly applied design rules can be detected across different user-interface domains and can be subjected to further research. Early analyses of the current set of over 600 image schema instances reveal

- rules of image schema co-occurrences (*e.g.* BLOCKAGE needs to be followed by RESTRAINT REMOVAL, ATTRACTION results in DIVERSION);
- image schema transformation rules (*e.g.* UP-DOWN is readily replaced by FRONT-BACK relations);
- typical problems (*e.g.* UI elements that belong to the same task are often far away from each other without communicating their relation via a link or common container image schema).

11.4 Image Schemas for Inclusive Design

Image schema theory holds a great promise for inclusive design. By keeping to a very basic level of prior knowledge it is assured that schemas will be accessible to the widest possible range of people. Because of their frequent encoding and retrieval from memory they are automatically and subconsciously accessible. The same holds true for metaphorical extensions, *i.e.* basic correlations in experience that have been generalized to abstract concepts. User interfaces that are built on the basis of image schemas and their metaphorical extensions should be more intuitive to use. Preliminary evidence for this stems from experimental research (Hurtienne and Blessing, 2007).

This paper focussed on the practical aspects of including image schemas in a user-centred-design process. It has been shown that they are useful in every step of a user-centred-design process: in analysing the context of use, in specifying requirements, in producing design solutions, and in the evaluation of user interfaces. Image schemas are particularly useful in bridging the gap between requirements and producing design solutions. As the example of the MATCHING image schema showed, designers can easily find solutions that effectively apply psychological principles without the need of a large background in perceptual psychology (in this case the visual pop-out effect supporting parallel visual search; Treisman and Gelade, 1980).

Image schemas provide a meta-language for abstractly describing what meaning user interface elements should transfer to the user. At the same time they do not restrict designers too much in specifying the look and feel of a user interface element (see also Hurtienne *et al.*, 2008).

As the explorative study has also shown, user-interface designs with image schemas can, in principle, enhance the pragmatic and hedonic qualities of a product as well as the product's attractiveness. However, more research has to be undertaken. In particular, carefully designed studies are needed that differentiate between the results of applying a conventional user-centred-design process without image schemas and applying a user-centred-design process augmented with image schemas. In doing this research, the effects of image schemas not only on the usability of the user interface but also on the design process itself (*e.g.* time required to find appropriate design solutions) should be investigated.

Finally, more specific research is needed into the application of image schema theory to the design of more inclusive technology, going beyond software used in a standard business environment. Its promise for designing inclusive user interfaces is high. Applying the theory leads to more intuitive user interfaces that preserve mental capacity for the task at hand, reduce the need for training and are applicable for a wide range of users. First steps have been made that confirm this potential. Inclusive design researchers are invited to go on from here.

11.5 References

Bias RG (1994) The pluralistic usability walkthrough: coordinated empathies. In: Nielsen J, Mack RL (eds.) Usability inspection methods. John Wiley and Sons, New York, NY, US

Blackler A (2006) Intuitive interaction with complex artefacts. Queensland University of Technology, Brisbane, Australia

BSI (2005) Design management systems – part 6: managing inclusive design – guide. British Standards Institution, London, UK

Clarkson J, Coleman R, Hosking I, Waller S (eds.) (2007) Inclusive design toolkit. Engineering Design Centre, University of Cambridge, Cambridge, UK

Gibson JJ (1979) The ecological approach to visual perception. Houghton Mifflin, Boston, MA, US

Hampe B (ed.) (2005) From perception to meaning. Image schemas in cognitive linguistics. Mouton de Gruyter, Berlin, Germany; New York, NY, US

Hassenzahl M, Burmester M, Koller F (2003) AttrakDiff: Ein Fragebogen zur Messung wahrgenommener hedonischer und pragmatischer Qualität. In: Ziegler J, Szwillus G (eds.) Mensch & Computer 2003. Interaktion in Bewegung. B.G. Teubner, Stuttgart, Leipzig, Germany

Hurtienne J, Blessing L (2007) Design for intuitive use – testing image schema theory for user interface design. In: Proceedings of the 16th International Conference on Engineering Design (ICED'07), Paris, France

Hurtienne J, Israel JH, Weber K (2008) Cooking up real world usiness applications combining physicality, digitality, and image schemas. In: Proceedings of Tangible and Embedded Interaction (TEI'08), Bonn, Germany

Johnson M (1987) The body in the mind: the bodily basis of meaning, imagination, and reason. University of Chicago Press, Chicago, IL, US

Keates S, Clarkson J (2003) Countering design exclusion. In: Clarkson J, Coleman R, Keates S, Lebbon C (eds.) Inclusive design: design for the whole population. Springer, London, UK

Koffka K (1935) Principles of gestalt psychology. Harcourt, Brace & Co., New York, NY, US

Naumann A, Hurtienne J, Israel JH, Mohs C, Kindsmüller MC, Meyer HA et al. (2007) Intuitive use of user interfaces: defining a vague concept. In: Harris D (ed.) Engineering psychology and cognitive ergonomics. Springer, Heidelberg, Germany

Story MF (1998) Maximizing usability: the principles of universal design. Assistive Technology, 10: 4–12

Treisman AM, Gelade G (1980) A feature-integration theory of attention. Cognitive Psychology, 12: 97–136

Chapter 12

Sustaining Autonomous Living for Older People Through Inclusive Strategies for Home Appliance Design

M. Baskinger and B. Hanington

12.1 An Opportunity for Design

As the baby-boomer generation (born in 1946–1964) in the US approaches mature adulthood, our society will recognise that it is becoming the oldest population in recorded history (Dychtwald, 1999). This reality presents three immediate opportunities for industrial design/product development of everyday objects and systems:

1. to synthesise the attributes, qualities, and values of an ageing population into viable product systems that address an older person's specific and changing needs;
2. to develop approaches to empower this specific user group/generation for sustained autonomous living;
3. to transparently embed assistive features to avoid stigmatising older people as a population marked by cognitive and physical deficiencies (Baskinger, 2007).

Major home appliances serve important roles in helping people address basic needs in their daily lives. The ubiquitous nature of these objects affords an opportunity for a socially responsive change in thinking by manufacturers to address real, human-centred issues of usability for this specific, rapidly increasing older population. Since the inherent multi-functionality and complexity of appliance interfaces can alienate older people who may lack the visual acuity, cognitive reasoning, or manual dexterity to operate them, the next obvious step for appliance manufacturers must be to develop an holistic inclusive strategy to allow for such disparities in abilities between younger and older populations. In this space, there is fertile ground for investigation into the specifics of product design and interaction design including: capabilities/functionality, materials, typographic information, iconography and visual support, mapping, physical dimensions, sequencing *etc*. Establishing parameters to accommodate older people can enable new product visual language and visual forms that can include and appeal to a broader, multi-generational audience (Baskinger, 2007). The project presented here represents a two year research investigation on visual language and product interaction by Carnegie Mellon's School of Design for General Electric Appliances. The project was led by professors Mark Baskinger and Bruce Hanington of the Industrial Design program with a small team of undergraduate and graduate design students for support. In the context of this chapter, this project is used as a case to illustrate key methods and approaches aimed at understanding the current older population and forecasting for baby-boomers as they progress towards their later life.

12.2 Understanding Needs to Clarify Design Opportunities

12.2.1 Building Empathy Through Experience and Observation

When designing for niche populations, it is important for designers to understand the user's lifestyle, behaviours, and the role that appliances play in their daily lives. To better understand appliance interaction from the target user's perspective, research team members first participated in "geriatric sensitivity training," provided by a gerontologist who consulted on the project. The goal of this was for the research team, aged 21–44 years, to build empathy for the physical and sensorial changes one experiences with ageing. During geriatric sensitivity training, physical and sensorial abilities of the research team members were limited using various devices such as glasses covered in Vasoline that distorted vision and thick latex gloves that reduced tactile sensitivity and dexterity. Team members were tasked to sort pills, thread needles, and read small type, under simulations of environmental lighting conditions that mirrored typical kitchen and bath contexts. Additional exercises included tasks completed with restricted mobility to simulate arthritis or stroke, and under conditions simulating hearing and taste deficits. This training was extremely important in preparing the group for visitation with older research participants and subjects, where the team would observe older people in their own homes using their appliances. Building empathy for older people enabled researchers to devise more insightful, yet respectful questions and tasks. Since the focus of the project was on empowerment and autonomy, the team needed to establish rich connections with participants and focused questioning and tasks on their abilities rather than pointing out their disabilities.

Field research was conducted with representatives from the older people and late baby-boomer populations in their own homes to best understand the daily challenges older people face in ordinary tasks. Field research or contextual inquiry, as it is commonly referred to, consisted of conversational interviews based on a framework of questions targeting past appliance experiences, current usage, and possible future needs, as well as noting individual preferences. Interviews and observations ranged from 45 minutes to two hours, which included a videotaped tour of appliances simulating typical use, while participants identified positive and negative features. In addition, telephone and email surveys were conducted with a smaller sample of specific members of these communities which followed a similar set of guiding questions (Baskinger, 2007).

Task analyses were completed with volunteer participants cooking meals for the research team in their home kitchens. Through task analysis, the research team learned that explicit issues raised in previous conversations were evidenced in practice. In addition, the team also found issues of product miscommunication which contributed to user mistakes. Multiple video cameras were installed in key locations in the kitchens to capture activity simultaneously from activity, context, and user perspectives. Ultimately, all views were synchronised and formatted into a video matrix to show what the participant was doing in relation to the context, the activity, and the appliances at every given moment (Figure 12.1). This form of observational study revealed that many participants made claims like "I know where everything is in my kitchen." or "I know exactly how my microwave works." However, recorded video revealed otherwise. When the video was played back to the participants, they were often surprised and quickly offered suggestions for how they might improve appliance interfaces and kitchen configurations to be more effective and efficient (Baskinger, 2007).

Figure 12.1. Task analysis through direct observation and video recording. This image shows the subject reading directions from a cookbook while food simmers on the cooktop and bread bakes in the oven.

12.2.2 Using Participatory Design Methods to Identify User Preferences and Needs

Including older people and boomers in the design process through generative modelling exercises, provided great insights of their preferred and desired appliance configurations. For participatory activities, a portable modelling kit was developed, targeting functional preferences for microwave and cooktop configurations. The construction kit contained card and paper templates with various appliance functions, displays and controls. Participants were asked to select preferred features, arrange them, and then simulate assigned cooking tasks (Figure 12.2). The exercise was informative in identifying preferred functions, sequences of common actions, and mapping of controls, setting some parameters for concept development. Conversations revealed a preference for controls to be located up front and easily accessible with larger knobs and more surface area for interfaces and visual feedback. Nobody seemed to prefer to reach over hot boiling pots to access control knobs. There were additional differences between the two populations; older people preferred fewer features while boomers preferred more features, including novelty features like one-touch popcorn buttons and baked potato settings for microwave ovens. Dialog with each participant reinforced the design team's assumptions that current older people prefer more control during food preparation, while boomers tend to rely on pre-set functions and automation (Baskinger, 2007).

Some key findings drawn from the exploratory and participatory studies included:

1. boomers tended to place more trust in their appliances and preferred pre-set options in dishwashers, microwave ovens, and washing machines;
2. older people expressed a desire to have more control during food preparation, laundry, and dishwashing, and tended to keep a watchful eye on appliances while in cycle;
3. older people and boomers had varying opinions on physical vs. digital controls that aligned with their acceptance of and proficiency with digital technology (Baskinger, 2007).

Figure 12.2. Participatory design with a baby-boomer participant using generative modelling for cooktop configuration. Note this participant prefers a more obvious mapping of controls close to the burners.

12.3 Product Development Strategy

12.3.1 Designing "Appropriateness" into Form Language

Appliances spend much of their life in a "ready" state waiting to be used; therefore, a strategy was developed to design the appliances to express their functionality and behaviour in a more overt way through movement and visual form. By reducing the visual "noise" (Baskinger, 2001) or visual complexity popular today in product forms, a more subdued presence was achieved to balance the visual presence of each appliance with the visual presence of other elements in context. The forms were envisioned to not dominate the visual landscape and detract from the interior design of their context – a key departure from contemporary strategies in appliance design where manufacturers try to call attention to their products in showrooms and homes (Baskinger, 2007). Developing a structured system of visual language rather than styling shifted thinking to the design of product experiences and promoted usability and interaction (Baskinger, 2001). This strategy required an understanding of how appliances are understood and used by current older people, to address their specific abilities, and to create appliance forms that resonated with them on more intuitive and emotional levels. The approach was to design clean-lined appliance faces with large digital information displays consistent across each appliance in the suite. This visual unity may help to promote confidence as interaction sequences can be expected and common visual language across each appliance can provide similar levels of feedback and information.

Simplicity of appearance and interaction drove form development so that the visual presence of each appliance expressed its functionality through the configuration of elements and the careful organisation of touch points for interaction. The flat panel display incorporates a physical knob with touch screen digital interaction similar to many contemporary automotive dashboard displays. The linear rectangular real estate enables digital information to be displayed through sequential visual organisation to create a pictographic story or narrative. The simple graphic language in conjunction with the simple product form give a "quiet" presence to each appliance in context which may enable most users to focus attention on the active interfaces and their own activities. Eliminating distraction and visual stressors within the appliance space emphasises the feedback on the information display which may promote confident and competency (Figure 12.9).

12.3.2 Establishing the Parameters: The Strike Zone

In the context of the GE project, the researchers focused on physical reach limits and optimal working spaces around each appliance, designing within current standards of American kitchen cabinet architecture. The "strike zone" concept was created with the goal of preserving the counter top as the major work surface in the kitchen and maximising access into and around each appliance. This concept delineates an optimal vertical area from 17" above the floor to 30" above the counter top, determined to minimise excessive stooping, bending and reaching.

Further, each appliance was configured to limit encroachment into the walking path in their primary open configuration to 13" which enables greater access for assistive mobility devices (Figure 12.3).

Optimal Vertical Workspace 17"- 63"
Strikezone 28" - 54"

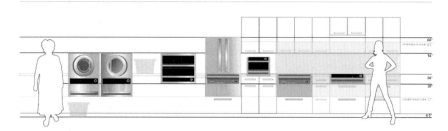

Figure 12.3. The "strike zone" concept illustrated in elevation with appliance suite organised linearly

Testing of volumetric models by representatives of a wide range of ages validated certain advantages that may improve appliance experiences for all. For example, the drawer dishwasher concept has a turntable dish rack that enables access from three different combinations of movement:

1. the drawer can be extended to full 13";
2. the lid can be lifted and stowed below the counter line to allow 13" of access from the top only;
3. the drawer can be extended 13" and lid opened fully to give 26" of access (Figure 12.4).

Most participants found the increased access at the counter line with multiple points of entry more preferable than current appliance configurations that promote excessive bending and stooping. Thus, an advantage created specifically to empower older people with poor balance and limited mobility provided a convenience for more able-bodied users (Baskinger, 2007).

12.3.3 Accessibility, Uniformity and Movement

This appliance suite includes configurations that share common carcass/shell construction, door fronts, and hardware as a modular platform. Unique appliance-specific features that promote access, readability and usability for various functions are built in accordingly. The goal in form development was to establish a unified appearance with consistent handles, knobs, controls and similarity in behaviour of moving parts. This approach greatly unifies interaction, simplifies manufacturing and establishes a consistency in the brand image. In addition, the counter line is emphasised through the positioning of appliance doors and features that promote sliding and minimise excessive lifting, bending and reaching (Figures 12.4 – 12.7).

Figure 12.4. Refrigerator: a counter depth fridge unit above the counter line, and a separate counter line beverage centre with turntable storage puts heavy gallons of milk and beverages close to the counter line to limit lifting/carrying distance. The beverage centre can convert to cold storage by removing the turntable and can also convert to a freezer unit.

Figure 12.5. Wall oven: a split folding door allows hot interior surfaces to mate, thereby minimising reach-over distance and accidental burns. The door opens at the counter line to provide a transition surface.

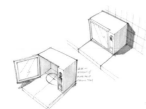

Figure 12.6. Microwave: a slide-in unit into the notched counter surface places the turntable at the counter line. This configuration enables older people with arthritic hands to slide hot dishes and liquids safely from within the microwave to the counter surface unlike contemporary microwaves where the lip of the turntable and transitioning to the counter top often causes spills.

Figure 12.7. Dishwasher: a 19" tall and 30" wide under-counter pullout drawer, with turntable dish rack and counter level hide-away lid. All dishes are accessible through the top lid or the drawer can be pulled out a few inches to gain further access but not encroach completely into the walk area.

12.3.4 Design Advantages

Five key areas summarise the design advantages of these appliance concepts validated by user feedback and participant survey: *Performance:* cooking, cleaning and storage tasks for the user are improved through this specific appliance suite configuration, location, visual form and behaviour. By positioning the primary activity areas at the counter line (strike zone concept), users are primed for more efficient performance with minimised bending and stooping. *Comfort:* larger, integrated grips and handles enhance control for hands with limited dexterity. The handles are integrated into the door face of each appliance so that bar ends do not protrude into the walk areas and therefore, pose no risk for accidental bumping or snagging loose clothes. *Ease of use:* larger access openings, the ability to slide cookware across surfaces onto the countertop, the strategic location of physical activity at the counter line provide a more beneficial scenario for older people (and everybody) where bending and stooping are reduced and awkward lifting of heavy or dangerously hot objects is minimised. *Safety:* handles at counter line can double as guide rails and grab bars should a slip or fall occur. Visual information displays show the complete cooking/cleaning process at a glance to enable the user to know where they stand in relation to the activity. Visual and auditory alerts become prominently displayed as "narratives" on the large information panels. Hot surfaces are reduced and controls are placed at optimal configurations to minimise accidental burns. *Access:* uniquely articulating doors and drawers as described above promote access to heavy, awkward objects. Wider openings allow for increased reach without hyper-extension of the limbs. (Baskinger, 2007).

12.4 Designing for Interaction

12.4.1 Narratives, Markers and Key Moments

Designing for key "moments" of appliance interaction creates touch points for the interactive sequencing to help clarify and define the cooking/cleaning experience. Moments, described as particular interactions and episodes within the greater product experience, can be isolated, categorised, designed and analysed (Figure 12.8), (Pine II *et al.*, 1999). In sequential structures, there must be a connection with past, present and future actions for the user/viewer to position themselves against. A narrative structure contains markers or cues set to initiate interaction and can answer such questions as: What can this appliance do? What can I do with this appliance? What must I do to engage it? With respect to illustrating the time continuum of the experience, this structure can also enable the user to formulate the following appliance-specific questions and provide concrete feedback about: What has the appliance done? What is the appliance doing now? What will the appliance do next? (Baskinger, 2007). Narrative structures can also help to address user-centred assessment during the interaction: What did I do? What must I do next? Since the abilities of the ageing population are extremely wide and varied, designing for key moments can help to clarify and inform interaction through a cause and effect relationship – *If I do this, then the machine will do that.* Like chapter headings of a book, key moments are the high-level touch points for interaction that can remind an older person of their position within the large process while simultaneously forecasting their next move. Scripting and designing specific moments into product interaction through cause and effect sequences can provide a more effective series of smaller interactions throughout (Baskinger, 2007).

Figure 12.8. An incomplete selection of photo images of some key moments in the laundry sequence with a current washing machine. The sequence starts with soiled laundry and progresses to the selection of was procedures. Key moments can be specifically designed for since they create valid touchpoints and markers during interaction.

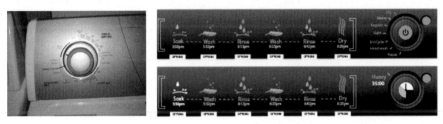

Figure 12.9. Contemporary washing machine interface example (left) showing all of its settings in small type. Narrative interface concept (right) developed for GE to illustrate wash cycles with simple pictographic animations to clearly communicate the various steps in the wash sequence.

12.4.2 Graphic User Interfaces to Promote Confidence

Graphic visual language has the ability to simplify, clarify and situate interaction. Within the GE project, the decision was made to commit to a large, graphic information display system that relied on pictographic representation to support and clarify text-based communication (Figure 12.9). Many concepts investigated adaptive interfaces that presented pertinent information only at decision points. For example, in many contemporary washing machine interfaces, all of the cycles and settings are displayed to a user all the time; even newer digital interfaces cram their screens with choices, pathways and settings that require translation. Thus, the research team made another commitment: to sequence physical and digital interaction in such a way that the first decision, generally the major cycle setting, was initiated in mechanical way through physical movement or tangible interaction; preferences and modifications to that setting would then be offered through digital touchscreen interaction as these would change following on from the first decision. Finally, the appliance would be actuated to begin its work cycle through physical interaction again. The combination of digital and tangible interaction elicited positive feedback from participants and provided a plausible approach for GE designers to make a greater commitment toward integrated digital interaction in future product platforms (Baskinger, 2007).

12.5 Key Findings

Organised feedback sessions with target older people and boomer user groups evaluating the design concepts in digital and volumetric mock-ups revealed that this conceptual suite of appliances begins to address some of the critical issues in empowering older people towards sustained autonomous living. Focusing on the communication and formal properties of each appliance to guide user interaction and shape behaviour enables these products to have a more prominent role in daily activities. While feedback continues to be gathered by the research team and GE designers, key take-aways from this project include the following:

1. design process and research methods can uncover a variety of hidden issues in product usage scenarios directly from user studies and participation;
2. product form language, not branding or styling, can be structured to express behaviours and shape user interaction;
3. narrative interface displays can inform user interaction and demystify appliance processes by presenting appropriate amounts of information and feedback at key moments;
4. developing inclusive product strategies can lead to new product forms and rich interaction that can benefit a much broader range of potential users.

12.6 References

Baskinger M (2001) Visual noise in product design: problems and solutions. In: Proceedings of the IDSA National Education Conference, Dulles, VA, US
Baskinger M (2007) Autonomy and the aging population: designing empowerment into home appliances. In: Proceedings of the 3rd European Workshop on Design and Semantics of Form and Movement (DeSForM 2007), Northumbria University, UK
Dychtwald K (1999) Age power: how the 21st Century will be ruled by the new old. Jeremy P. Tarcher/Putnam, New York, NY, US
Pine II BJ, Gilmore JH (1999) The experience economy. Harvard Business School Press, Boston, MA, US

Part III

Computer Access and New Technologies

Chapter 13

Investigating the Security-related Challenges of Blind Users on the Web

J. Holman, J. Lazar and J. Feng

13.1 Introduction

Historically, there has always been a trade-off between usability and security. In addition, increased security also tends to increase the complexity of interfaces for users with disabilities. This research paper has three important components: 1) a focus group with blind users to determine their top 10 security-related challenges, 2) development of a new prototype to address one of the challenges for blind users, and 3) evaluation of the prototype with five visual users and five blind users. The prototype developed is a new form of CAPTCHA, based on the combination of non-textual pictures and sound clips, which is equally usable by both visual and blind users. Data collected shows that this new form of CAPTCHA is preferred by both blind and visual users to previous forms of text-based CAPTCHAs. Future directions for research are also discussed.

There are many studies of usability and security, and there are many studies of usability and users with disabilities, but research at the intersection of these two topics is rare. Specifically, there is not a great understanding of what security-related issues pose a challenge to usability for blind users on the web. The goal of this study is to fill this gap.

This project includes three important and inter-related components: 1) a focus group to develop a "top 10 list" of security problems for blind users, 2) development of a prototype of a new, more accessible, more secure form of CAPTCHA, and 3) evaluation of the new CAPTCHA with both blind and visual users.

13.2 Literature Review

In the world of information security, there is always a usability/security trade-off. The more secure you make an interface, the less usable it becomes (Sasse *et al.*, 2001; Johnston *et al.*, 2003). The more usable you make it, the less secure it is.

This becomes even more of an issue when you are trying to make systems usable for users with disabilities. For instance, blind users rely on screen readers to use their computers. A screen reader is a software package that produces synthesized speech output for users, based on the text on the screen and other back-end code. Screen readers introduce some very specific usability requirements. When designing web sites, security is also a very important issue. While there is not a great base of literature on security and blind users, there are a few references on security and users with disabilities in general. For instance, D'Arcy and Feng (2006) found that users with motor impairments use simple, non-complex passwords that are usually short in length due to their difficulty in using the keyboard. This is obviously a security concern, as these account passwords will be much easier for an intruder to crack.

According to W3C (2007), CAPTCHAs were one of the greatest security-related problems for users with disabilities, especially blind users. CAPTCHA stands for Completely Automated Public Turing tests to tell Computers and Humans Apart. This is based on the test developed by Alan Turing in the 1950s to test a machine's ability to imitate a human. Usually these CAPTCHAs are a visually garbled image of a set of letters and numbers that humans can identify, but a computer application would not be able to. The reason for garbling and distorting the image is that if it were just plain text in an image file, it would be very easy for Optical Character Recognition (OCR) software to identify it. So researchers garble the image to take advantage of human error-correction abilities. In 1997 AltaVista came up with a CAPTCHA like concept to protect its search engine from having web sites submitted by bots (Robinson, 2002). The concept was later expanded upon by researchers at Carnegie-Mellon. Since their creation, CAPTCHAs have been successful for the most part, in serving the purpose of creating and grading tests that a majority of humans can pass, but automated computer programs can't pass (von Ahn, 2003). Another use of CAPTCHA is to prevent dictionary attacks. A dictionary attack is when a program tries to break into an account by guessing passwords using words in the dictionary. Gmail makes use of this strategy; if you enter your password incorrectly three times, your next login requires your correct password and a CAPTCHA.

CAPTCHAs have extremely valid purposes; they are built to keep web bots from filling out and submitting forms. These web bots usually perform tasks like this to obtain free email addresses for spammers to use to send their spam or to submit spam to online blogs and forums. There does need to be some sort of mechanism to prevent these illegitimate users from submitting forms. Unfortunately the current CAPTCHA method used blocks out users who cannot see the twisted text because their screen reader accessibility software has no ability to discern it. Of course, if the screen readers could interpret the CAPTCHAs, so could the web bots. So the current "distorted text" CAPTCHAs continue to be problematic for blind users. There are audio CAPTCHAs, which provide an audio version of the letters and numbers on the screen for blind users, but there must be a significant level of background noise in the audio clip, otherwise voice recognition software could easily identify the text in the audio clip and it would become less secure. Partially for that reason, audio CAPTCHAs are rare on web sites.

13.3 Requirements Gathering Through Focus Group

During the first stage of this study, a focus group was held at the National Federation of the Blind in Baltimore, Maryland, moderated by two of the authors of this paper. The goal of this focus group was to learn more about the security-related problems of blind users. The group consisted of employees of the National Federation of the Blind, all of whom are legally blind and use screen access software and/or Braille displays as their primary approaches to interact with computers. The group resulted in a "top 10 list" of security related problems for blind users on the web.

1. Web sites with forms using visual CAPTCHA and no audio version to authenticate users are inaccessible to blind users. Since the blind users cannot read the visual CAPTCHA without assistance, they are required either to ask a colleague for assistance or phone tech support. PayPal and AIM both have audio CAPTCHAs that work, so it is suggested that more companies imitate this, if they must use a CAPTCHA.

2. Many web sites have secure login sessions that time-out and log the user out automatically after a few minutes of inactivity. Unfortunately, some users, especially users of assistive technologies, can sometimes take longer to fill out the forms. So when the web site thinks that the user is inactive and has perhaps walked away from their PC, the user is actually still filling out the form. The user is then automatically logged out, forcing them to log back in and start the form over.

3. Some web sites occasionally automatically refresh/reload the page's content. While there might be valid security-related reasons for this, when it happens it causes the screen reader to start over in reading the page, as if the user had just opened the page for the first time. At the same time the user probably isn't aware of what is going on and why, because they did not request the page to be reloaded. Web pages should only reload per the user's request.

4. While PDF files on web sites can be made accessible, often they are not accessible. When creating an Adobe PDF file you have the option to allow changes, or "lock it" and not allow changes. If you allow changes, any users can copy and paste, and make modifications to the document. If you lock the document so that changes are not allowed, you must specifically note that you want the text to be accessible to users with assistive technology. If this accessibility option is not selected, and the document is locked, then the PDF file may be inaccessible to screen reader users. Inaccessible PDF files are often a result of PDF files that were created a while back, by using much older versions of the PDF writer.

5. Some of the latest antivirus packages are still inaccessible. Norton 2007 is apparently inaccessible, although 2006 was very good. There are also some versions of McAfee antivirus that are inaccessible. This is obviously a security concern as blind users want to protect themselves from viruses, but are unable to do so using some of the most current versions. It's another trade-off: keep the older virus software, which is accessible, or upgrade to the newer virus software, which is more secure but less accessible.

6. Often when a user loads a web page, some sort of software tries to automatically install. Usually this is spyware and it should not be installed. However, many blind users are not aware of what is attempting to install and may not be presented with enough information to allow them to make an informed decision about installation. Improvements must be made to provide the users with more information about what piece of code is attempting to be installed.

7. Operating system and application updates can sometimes make the software inaccessible. This is another security compromise for users: if they update they could lose accessibility, but if they don't upgrade they could compromise security. Because of this, many blind users have disabled automatic updating of software applications.

8. Another concern is the potential use of SecureID. SecureID is a handheld device that displays a number: this number changes every couple of minutes and when you want to log into a secure site or your company's VPN (virtual private network) you type in this dynamic number plus your PIN. This allows you to log in and provides a more secure key than your PIN or the dynamic number alone. Obviously a blind user would be unable to read the number displayed on the SecureID device, so whatever system is being secured by this means would be inaccessible to them.

9. Key loggers are a malicious piece of software that can be installed that logs every key that the user presses. These logs can later be reviewed to disclose an unknowing user's password, credit card numbers, or other private information. These pieces of malicious software could be even more of a threat to blind users because of their dependency on the keyboard (and non-use of pointing devices).

10. Spam is very annoying and inconvenient to all users, including blind users. Since blind users need to listen to the text of their e-mails, it may take longer to discern the true nature of the e-mails (SPAM!) than a quick visual scan of an e-mail would take. It can also cause many users, especially underage users, to be caught off guard by the content of some of these unsolicited junk emails.

13.4 New Form of CAPTCHAs

Following the discussions in the focus group, it seemed that the problem of CAPTCHAs was one which was increasingly important and needed effective alternatives. We decided to create a prototype of a new form of CAPTCHA, which could accomplish the same goal of user verification, but be easier to use, not only for blind users, but also for other users. Our idea was to create a CAPTCHA that did not use words, but instead used corresponding pictures and sounds that presented the same concept. The pictures would make the CAPTCHA usable by someone who could see, and the sound would make the CAPTCHA usable by someone who could not see. This combination of sound and images should make the CAPTCHA more universally usable for a large majority of users. In addition, the use of concepts, rather than text, makes the CAPTCHA, at least for now, more secure, as image processing and voice recognition are not at the point where they can easily identify concepts. Another benefit of this form of CAPTCHA is that it could be multi-lingual without much work. Since the new CAPTCHA would use pictures and sound effects, many of these concepts (although not culturally-specific ones) could be used all over the world. The only thing that would need to be changed to develop the system for another language is the labels for the sound/image combinations.

The prototype tool is a web-based application. Upon loading, the web page displays a picture and offers a button to load a corresponding sound clip. We chose four categories of picture/sound combinations: transportation, animals, weather, and musical instruments. These four categories were chosen because they were easy to recognize for a majority of potential users, without any special training or expertise. The researchers spent time finding appropriate clips, and all researchers had to agree on the identification before they could be included in the application. One example is an image of a train and the sound clip of a train chugging along. Other examples included bird, cat, drum, and piano. Any items that had multiple easily identifiable labels were discarded. The web-based application was tested in both JAWS^{TM} and Window-Eyes^{TM} to make sure that it was fully accessible with the most common screen readers. When the user views the image or listens to the sound clip, they are prompted to choose the correct label from a drop-down list that also includes many items which do not have images or sound clips. Figure 13.1 shows a screen capture of the user interface of the prototype application.

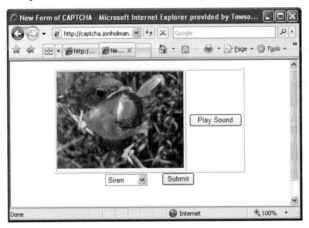

Figure 13.1. Screen shot of the prototype CAPTCHA interface

13.5 Usability Testing

In order to evaluate the usability of the new CAPTCHA, a small scale proof-of-concept user study was conducted involving both blind users and visual users. For this testing, the tool was modified to go through objects one through 15 sequentially, so that the user who was testing the tool would not see or hear the same object twice.

13.5.1 Visual Users

Five visual users tested the new CAPTCHA prototype. Each user went through the 15 CAPTCHAs, identifying the title of each object from the list. After testing, we had the participants fill out a survey to gather their thoughts on the new tool. All of the participants completed the 15 tests within 3-5 minutes (average: 3.5 minutes) with no problems, identifying the objects that the pictures and sounds represented. No errors were made by any of the participants. Although the visual users had the benefit of using both the picture and sound to identify the object, our goal in designing this tool is for the picture or sound to be sufficient, on its own, to identify the object.

Table 13.1 summarizes the answers to the survey questions from both visual users and blind users. Note that while the rating of '1' represents the most positive response in most questions, in other questions, the rating of '1' represents the most negative response. In this way, the participants would be more likely to read the questions carefully instead of marking either '1's or '5's for all questions without even reading them.

The visual users found it very easy to accept the design idea of using both images and sounds to identify objects (1.2 out of a scale of five, one being the easiest to accept). Overall the visual users were highly satisfied with the ease of use of the application and the time spent to complete the tasks. They had no difficulty identifying the objects based on images and sounds. Encouragingly, although the visual users can effectively use both the traditional and new form of CAPTCHAs, the survey showed that they somewhat prefer the new form of CAPTCHA (Average 3.6 out of a scale of five).

Table 13.1. Summary of survey results from both visual and blind users

Questions	Visual users	Blind users
Overall ease of use (1=easiest to use)	1.2	1.4
Acceptability of design concept (1=most acceptable)	1.2	1.4
Satisfaction with time (1=most satisfied)	1.6	1.2
Difficulty in identifying object (1=no difficulty)	1.4	3.2
Preference of sound over image (1=prefer sound)	3.2	1
Preference of text-based CAPTCHA over prototype (1=prefer text-based CAPTCHA)	3.6	4.8
Likelihood of visiting a site using the prototype (1=very likely)	1.8	1

13.5.2 Blind Users

Five blind users were recruited to test the prototype. All five blind users interact with computers or computer related devices via screen reader or Braille keyboards in their daily life. The blind users tested the same version of the CAPTCHA prototype and responded to the same survey as the visual users.

The blind users took on average 8.8 minutes to complete all 15 questions. Two users got through all 15 objects in a little under eight minutes and the longest test was 17 minutes (which was done with a slower screen reader reading speed, at the user's preference). The user who took 17 minutes was really exploring the application, and gave a lot of verbal feedback. So time cannot be the only metric used to measure performance.

Regarding accuracy, the average error rate was 9.3% (stdev: 0.036). Three of the blind participants made one error during the test. Two blind participants made two errors. On the second attempt, all users were able to choose the correct choice from the list of items. Since most security-related features allow three attempts, we feel that this was a successful prototype, since the error rate in three attempts would be zero.

According to the survey results shown in Table 13.1, the blind users, the same as visual users, also found it very easy to accept the design idea of using the new form of CAPTCHA (1.4 out of a scale of five, one being the easiest to accept). They were highly satisfied with the ease of use of the application and the time spent to complete the tasks. The blind participants did not find it difficult to identify the object based on the sound provided (Average 3.2 out of five, three being neutral).

Regarding the possible adoption of the new form of CAPTCHA, the blind users greatly prefer this new prototype over the traditional forms of CAPTCHA (average 4.8 out of five, five being highly prefer the new CAPTCHA) and expressed that they would like to visit sites with the new form of CAPTCHA (Average one, with one being most likely to visit), as they are unable to use the majority of current text-based CAPTCHAs. Overall their survey results showed that they were highly satisfied with the proposed new form of CAPTCHA.

13.6 Discussions

13.6.1 Selection of Sound Effects

The selection of appropriate matching sound clips and images is a crucial factor for user performance. The match between the term, the image, and the sound effect should be clear, unique, and obvious. Sometimes multiple concepts/terms may be connected to the same object. For instance, we had both thunder and lightning as valid answers for the sound of thunder and the image of lightning. We also had both alarm and siren as valid answers on the sound of a siren. Even with those pre-cautions, a few sounds proved to be problematic during the actual test. For instance, it was found that the pig sound effect was not clear and obvious enough.

With visual users, they could look at a picture of a pig, or see the picture and also listen to the sound of a pig, which made it more obvious. However, the object presented a problem for blind users since the pig sound effect was of rather quiet grunting, and did not contain the typical 'oinking' associated with pig sound effects. In another test, one participant was looking for the word fox when the wolf howling was played. The sound effects that were well received by the participants included glass, truck, train, siren, and bell.

13.6.2 Presentation of Sound Effects

The sound effects need to be a certain length to be clear to the users. A large part of this is caused by the screen reader software because with this every key that you press is spoken through the computer speakers. So if you use the enter or spacebar key to press the "Play Sound" button, the computer will be saying "Enter" or "Space" while the sound is playing. If the sound is not long enough to keep playing after the screen reader feedback, the user will not be able to hear it clearly. This showed that it might be necessary to repeat the sound effects a few times. For instance, the cat sound clip was very brief and had a cat meow only once. A few users missed the sound the first time they heard it. In contrast the dog sound has a dog bark three times, which was easier for the users to capture. If the user wasn't sure what that was the first time, the repetition helped out. Another suggestion to compensate for the screen reader software reading the key presses was to insert a delay before the sound plays. The cost is that this will slow down the time it takes for users to complete the CAPTCHA test and that a delay on a web site can give the impression of a poor server, slow connection, or a web site that is currently down.

13.6.3 Interaction Strategies

Different strategies were used by the blind users to complete the CAPTCHA tests. Some users were able to go through the majority of the CAPTCHA tests very quickly. Their main strategy was to recognize the sound, go to the drop-down list, press the key for the first letter of that word, and quickly jump to the answer. Some other users heard the sound and then went through the list considering each item until they found what they believed was the correct answer. Other users went through the entire list on the first few CAPTCHAs, and then on the future tests they jumped to the letter using the first strategy. This new form of CAPTCHA supported multiple strategies for completing the task goal, all of them leading to success.

13.6.4 Considerations for Practical Application

We believe that this new form of CAPTCHA could be used by companies to help protect their web sites against automated attacks, while at the same time allowing effective use by individuals with visual impairment. If this new form of CAPTCHA is placed into practical use to secure websites against bots, there are a few things that could be done to improve the security in the long-run.

As pointed out in the literature about CAPTCHAs, it is very important for all CAPTCHAs to use an infinite database, or at least have a large set of options. This is because if there is a small and finite number of options, the bots could eventually index all of the challenges and know how to properly respond to each challenge. The major weakness of the new form of CAPTCHA is that it does not include an infinite number of options, due to the way in which audio and image files must be retrieved, paired, and then inserted into the application. So for our application, we suggest that an administrator should occasionally modify the database of images and sound files, so that a bot cannot catalogue the options.

Another feature that would be desirable is randomizing the file names on every run, or to having all file names randomly chosen renamed to temp before being displayed. Both of these will make it impossible for a bot to catalogue the filenames so that it will know how to respond correctly. This feature also has the complication of serving many users simultaneously, so it is important to make sure that each user's session files do not interfere with other concurrent sessions.

Another method that could be useful in preventing potential bots from cataloguing the challenges they are faced with is making the files look identical to a computer. This can be done by making all of the file sizes the same so they cannot be catalogued to determine what the answer is. Finally, as with current CAPTCHAs, if there are more than two or three incorrect responses from the user, the user should be locked out from further attempts.

13.7 Summary

This study investigated several important issues regarding security and usability for blind users. The top security concerns of blind users on the web were identified. A prototype of a new form of CAPTCHA, based on non-textual images and sound clips, was developed. A preliminary user study showed that this new form of CAPTCHA provided satisfactory interaction experience for both visual and blind users. It is particularly helpful for blind users since it addresses the problems of the traditional visual text-based CAPTCHAs and provides a secure solution that is significantly more accessible.

13.8 Acknowledgements

We appreciate the assistance of John D'Arcy on the development of the CAPTCHA prototype.

13.9 References

D'Arcy J, Feng J (2006) Investigating security-related behaviors among computer users with motor impairments. In: Poster abstract at Symposium On Usable Privacy and Security (SOUPS 06), Pittsburgh, PA, US. Available at: http://cups.cs.cmu.edu/soups/2006/posters/darcy-poster_abstract.pdf (Accessed on 10 March 2007)

Johnston J, Eloff J, Labuschagne L (2003) Security and human computer interfaces. Computers and Security, 22(8): 675–684

Robinson S (2002) Human or computer? Take this test. Available at: http://query.nytimes.com/gst/fullpage.html?res=9907E5DF163AF933A25751C1A9649C8B63 (Accessed on 17 March 2007)

Sasse MA, Brostoff S, Weirich D (2001) Transforming the weakest link – a human/computer interaction approach to usable and effective security. BT Technology Journal, 19(3): 122–130

von Ahn L, Blum M, Hopper N, Langford J (2003) CAPTCHA: using hard AI problems for security. Available at: www.cs.cmu.edu/ ~biglou/captcha_crypt.pdf (Accessed on 3 April 2007)

(W3C) (2007) Inaccessibility of CAPTCHA. World Wide Web Consortium. Available at: www.w3.org/TR/turingtest/ (Accessed on 10 March 2007)

Chapter 14

Access Barriers to Wireless Technologies for People with Disabilities: Issues, Opportunities and Policy Options

P.M.A. Baker and N.W. Moon

14.1 Introduction

While the adoption of wireless technologies in the United States continues to become increasingly widespread, significant issues of access to these technologies persist for people with disabilities. In the U.S., more than 51.2 million people, constituting about 18 percent of the population, have some kind of long-term condition or disability, signalling that barriers to the adoption of wireless technologies affect a substantial population (U.S. Census Bureau, 2006). Equal access to technology-related services and devices and wireless accessibility issues can be addressed by legislation and regulations, as well as options developed from disability and telecommunications policy and research.

This paper discusses the relationship between policy research and policy change, and also examples of the policy change outcomes of the Rehabilitation Engineering Research Center on Mobile Wireless Technologies for People with Disabilities (Wireless RERC, 2003) policy research process. In 2005 and 2006, the centre conducted empirical research, using policy Delphi polling methodology, to probe key stakeholders' opinions on the most significant issues surrounding the adoption and use of wireless communication and information technologies by people with disabilities. Drawing on the results of three rounds of polling, the Wireless RERC developed a set of policy options, and "fine-tuned" them using participating stakeholders from the disability community, wireless industry, and policymakers.

14.2 Policy Research and the Policy Change Process

The policy assessment and change process utilized by the Wireless RERC is grounded in 1) ongoing monitoring and evaluation of federal regulations and policy; 2) concomitant assessment of the impact of public policy, such as the Federal Communications Commission's (FCC) disability access rules and emergency alerting requirements; 3) consultation with pertinent key stakeholders; and 4) development of subsequent advisory materials, policy guidance, and filings to address barriers to use of wireless technologies by people with disabilities. These research findings are used iteratively to refine and identify the Centre's future policy research agenda (Baker and Bellordre 2003, Baker *et al.*, 2006, 2007).

The research focus of the Centre includes: evaluation of the impact of public policy, such as the FCC hearing aid compatibility regulations for wireless phones; development of standards encouraging "compatible platforms between wireless and other mobile devices used by people with disabilities;" and development of programs to encourage manufacturers of wireless devices to include people with disabilities in the review and evaluation of assistive/universally designed products. The policy change component of the RERC's activities manifests in consultation with stakeholders on activities of interest, production of informative newsletters and papers, generation of filings before the FCC, and other pertinent agencies, and in research collaboration in other, related, venues.

14.3 Methodology

The data for our findings was provided through a three-round electronic PolicyDelphi (e-Delphi) method, derived from the original Delphi method developed by Olaf Helmer and Norman Dalkey at the Rand Corporation during the 1950s and 1960s (Dalkey 1969, Dalkey *et al.*, 1970). The Delphi method was originally conceived as a tool for military and economic forecasting based upon iterative surveys of experts in the given area under consideration (Cornish, 1977). Modern Delphi method relies upon expert opinion, professional experience, intuition and tacit knowledge in order to render a forecast on a given issue of importance.

A policy Delphi seeks to develop pro and con arguments about policy issues and their resolutions (Turoff, 1970), and allows a panel of experts to contribute elements to a complex situation with the intention of building a composite situational model. Turoff conceived of a policy Delphi as "less about the use of experts to generate a policy decision, and more about employing a group of "advocates and referees" to present all the options and supporting evidence for a given issue, and "generate the strongest possible opposing views on the potential resolutions of a major policy issue" (Linstone and Turoff, 2002).

Policy Delphis have any of three important objectives: 1) to ensure that all possible options have been proposed for consideration; 2) to estimate the impact and consequences of any particular option; and 3) to examine and estimate the acceptability of any particular option. As conducted, the Wireless RERC policy Delphi considers possible options to increase use of and access to wireless technologies for persons with disabilities. The most important objective was to consider the feasibility and acceptability of the options proposed. Policy Delphis rely upon six phases in the communication process between their participants: 1) formulation of the issues; 2) exposing the options; 3) determining initial positions on the issues; 4) exploring and obtaining the reasons for disagreements; 5) evaluating the underlying reasons; and 6) re-evaluating the options (Linstone and Turoff, 2002).

Policy Delphis adhere to four key principles: anonymity, which minimizes outside influences on the predictions panellists make; asynchronicity, the ability of participants to take part when and how they choose to; controlled feedback, as the results of one round of questions are used to inform the creation of the next; and statistical response, taking the opinions of experts on a given area and converting them into quantitative data. The current e-Delphi was conducted via the Human-Environmental Observatory's (HERO) e-Delphi system, hosted by Pennsylvania State University.

In the case of the subject policy Delphi, an initial set of issues and goals were formulated prior to the first round through a review of the literature and the development of an analytic policy matrix. During the first two rounds of the Delphi, these issues and goals were presented to the panel for review. Open-ended responses were used to help formulate additional issues and goals through the Delphi, and to gauge reasons for disagreement whenever a proposed issue or goal failed to receive a strong majority of support. The third, and final, round of the Delphi was committed solely to a consideration of the feasibility of options.

The Wireless RERC Policy Delphi, conducted between October and November 2004 (Round 1), June and July 2005 (Round 2), and February and May 2006 (Round 3) probed key stakeholders' (N=70) opinions of the most significant issues for the adoption and use of mobile wireless technologies by individuals with disabilities. Participating stakeholders included members of the disability community, and the wireless industry, and policymakers. The instrument asked participants to assess the reliability of forecasts, importance of issues, desirability of goals, and feasibility of proposed options, in four areas: access and awareness, economic, regulatory and policy, and technology. In the third, and final, round of the Delphi, participants were presented with a set of detailed options to be assessed with respect to their policy feasibility. The options were presented as specific actions to be taken providing enough flexibility within each proposal to mitigate potential differences of opinion between stakeholders.

14.4 Overall Results

In the first round of the policy Delphi, three findings distinguished themselves in particular: 1) problems with device compatibility; 2) ongoing awareness issues, especially on the part of manufacturers; and 3) economic concerns focusing less on the affordability of devices, but rather more on the general level of employment of people with disabilities. The second round of the policy Delphi sought to elaborate on some of the key issues and goals established in the first round, and established some of the primary issues pertaining to manufacturer and designer awareness of disability issues. Participants agreed that while manufacturers fail to design appropriately for people with disabilities, a much larger divide exists between designers/manufacturers, consumers/users, and retailers/intermediaries. Manufacturers are often oblivious to the barriers which face consumers with disabilities, but likewise, potential consumers of wireless products lack information necessary to select and use such technologies.

In the third, and final, round of the Wireless RERC's policy Delphi, participating stakeholders were presented with the results of the first two rounds. Delphi participants were also informed about which questions elicited a consensus among respondents (as in the forecasts) and which ones resulted in discernable blocs of opposition (regulatory and economic goals, for instance).

14.4.1 Policy Options

Access/Awareness Options

The third round of the Delphi asked respondents to consider the feasibility of five options related to access and awareness issues regarding wireless technologies for people with disabilities. The access/awareness options presented were among the most strongly supported in this final round of the Delphi. However, in the comments, respondents occasionally voiced concerns that the presented options should be disaggregated to reflect better the feasibility of certain aspects over others, and that while certain options might be feasible as policy, they were of uncertain effectiveness.

First, Delphi participants were presented with an option to increase investment in public information campaigns about the availability, benefits, and use of wireless devices for people with disabilities. Such activities might involve the development of programs through the FCC Consumer and Governmental Affairs Bureau's Consumer Affairs and Outreach Division, at the federal level, as well as through State Assistive Technology Programs. Some participants noted that the FCC might not have the resources to become involved in such options and, without technical staff, could not be involved in the process of product evaluation. Reflecting concerns about the ability of federal and state governments to be involved in such an option, some respondents suggested that the involvement of consumer groups might be more effective. Second, participating stakeholders were asked about an option to launch campaigns to educate manufacturers of wireless devices about the economic viability of universally designed products, existing markets of people with disabilities, and perhaps most important, larger untapped markets of aged individuals. Industry groups could develop an internal promotional campaign aimed at mass-market manufacturers and other non-niche marketers of wireless technologies emphasizing a voluntary, market-oriented approach with an orientation toward outreach and education. Those in favour of this option contend that it remains important to raise awareness throughout the wireless industry, among manufacturers, service providers, and retailers, that "accessibility issues are important, some solutions are available, and accessible products will expand sales."

Next, the feasibility of training and education programs for educating retailers about product accessibility features was considered. Such programs would be led by manufacturers of wireless technologies for people with disabilities or their respective trade associations, such as the Assistive Technology Industry Association (ATIA) or the Cellular Telecommunications and Internet Association (CTIA).

The fourth option was to develop user forums and "demo rooms" where consumers with disabilities can review wireless products and provide evaluations in a range of specialties. Some participants noted that user forums and "demo rooms" could provide industry with a great opportunity to gather considerable amounts of consumer input; others stressed their win-win potential for industry and consumers with disabilities. Critics of the option noted that wireless technologies change so rapidly that it might be impossible to keep demo rooms and their staff up-to-date.

Participants considered the feasibility of a Consumer Reports-styled guide that would provide consumers with information about the usefulness and features of such technologies. Such a publication, the option suggested, might be developed and published by disability advocacy groups

Economic Options

Participants were decidedly split over the feasibility of the proposed economic options. While there was some support for the expansion of tax incentive programs to help employ persons with disabilities and promote the use of wireless technologies by them, there was also considerable disagreement over the expansion of existing equipment distribution programs and increasing income caps for access to wireless technologies by persons with disabilities.

First, participants deliberated over the expansion of state equipment distribution programs (EDPs), which generally follow one of two models: a voucher program that provides users with grants that may be used to purchase equipment from EDP approved suppliers, or a loan program that either provides clients with pre-purchased devices loaned on an as-needed basis or direct loans to obtain equipment. The FCC's Disability Rights Office (DRO) might also play a supporting role in such efforts. Some participants noted that the FCC lacks jurisdiction over such programs, and shows little interest in developing federal standards for equipment distribution. Second, respondents note the presence of many budget problems at the state level that impede the expansion of such programs.

The Delphi also probed the feasibility of expanding federal and state level initiatives to provide employers with additional motivations to employ people with disabilities, and for increasing tax incentives at the state and federal levels to promote the use of wireless technologies by people with disabilities. A majority, 68 percent, responded that the option was feasible while 23 percent of the Delphi participants believed such programs were possibly unfeasible, and an additional 9 percent declared them definitely unfeasible. Critics noted that currently available tax incentives, such as those for architectural and transportation barrier removal, have gone largely unused by employers. Others questioned whether policy options to increase employment of persons with disabilities and improve their socioeconomic status more generally might realistically promote the use of wireless technologies by persons with disabilities, arguing that accessibility of the hardware is the main barrier to their use, not the ability to purchase them.

Finally, participants considered the feasibility of a policy option to raise income caps imposed by assistance and distribution programs of such technologies. Given the greater expense of wireless technologies, this option asked whether increasing income caps to higher levels might enable greater access to wireless devices for persons with disabilities. A majority viewed this initiative as possibly unfeasible (57 percent). Of the respondents who regarded increases in income caps as a feasible option, many noted that while wireless technologies might be less expensive than the assistive technologies covered under many equipment distribution programs, the higher recurring costs of wireless technologies warrant serious consideration for this economic option.

Regulatory/Policy Options

Participants in the Delphi study deliberated on the feasibility of two proposed regulatory/policy options: 1) a regulatory enforcement study to determine whether current legislation and rulemaking has increased access to wireless technologies by persons with disabilities; and 2) programs designed to strengthen the relationship between public sector and private sector research and development of wireless technologies that would benefit persons with disabilities.

First, respondents were asked about the value of developing a study or survey to determine whether the enforcement of Section 255 of the 1996 Telecommunications Act and Section 508 of the Rehabilitation Act of 1973 has been effective and resulted in greater access to and accessibility of wireless technologies for people with disabilities. This option developed from a consensus among participating stakeholders that too little is known about the impact these regulations have on mediating barriers faced by persons with disabilities. Support for the feasibility of this option was among the highest of those presented. 86 percent of participants found such a study or survey would be feasible.

Second, participating stakeholders were asked to consider the feasibility of increasing links between public sector and private sector research and development of wireless technologies of people with disabilities. In addition, trade associations might also play a role in strengthening relationships between public sector and private sector research. A majority of participants, 90 percent, agreed on the feasibility of this proposed option. Many participants noted that this goal of improving relationships between the private and public sectors should be a central part of the RERC's missions and some noted that a few of the RERC's are currently engaged in such initiatives. Some respondents in favour of the initiative noted that inadequate funding might pose a problem to such a venture.

Technology Options

Finally, respondents were asked to evaluate the feasibility of the development of new interoperability and technology standards, and the feasibility of developing initiatives through the FCC Enforcement Bureau's Emergency Alert System (EAS). Delphi participants were first asked about the feasibility of developing a voluntary set of standards for product interoperability, compatibility, and accessibility for users with disabilities, with manufacturer trade associations working in collaboration with federal agencies such as NIDDR and the FCC. Some respondents questioned making standards voluntary, noting that Section 508 provided a better model. Second, some participants contended that it would be better to work with existing standards and simply add accessibility to them. A third viewpoint predicted manufacturers' concern about disclosure of proprietary processes.

Second, Delphi participants assessed the feasibility of developing initiatives through the FCC EAS to develop device guidelines and protocols to receive and recognize EAS alerts. This option also suggested the involvement of disability stakeholders to provide FCC advice for rulemaking, and received the most support of all of those presented to the Delphi respondents. An overwhelming majority, 94 percent, believed it to be feasible and noted that such policies were likely to have broad public support that could help in building consensus among the multiple conflicting interests of stakeholders.

14.4.2 Comments and Analysis

At the outset of the Delphi, there was pronounced emphasis on economic and technological barriers to the adoption of wireless technologies by persons with disabilities. However, during the first two rounds, it became apparent that access/awareness issues warranted more attention. Moreover, whereas emphasis was placed early in the Delphi on awareness of wireless technologies by consumers with disabilities, equally important was directing the awareness of designers and manufacturers to the needs of persons with disabilities. The fact that participating stakeholders in the Delphi agreed on the feasibility and importance of five policy options related to access and awareness, more than any other issue, speaks volumes about their salience in addressing adoption barriers.

The Delphi also revealed a number of tensions inherent in the process of addressing wireless technologies and persons with disabilities. First, the Delphi instrument raised the issue of whether the policy options generated in the course of the Delphi should be framed as mandates or as voluntary or market-based initiatives. Respondents expressed strong support for initiatives that were either voluntary or collaborative efforts, such as developing manufacturer or designer standards. Conversely, there was little support for policy mandates regarding the accessibility of wireless technologies by persons with disabilities. Delphi participants suggested improving the understanding of manufacturer, designer, and retailer awareness about the existence of markets for such technologies and how the needs of various disability groups could be taken into account when designing and marketing relevant products. Likewise, economic options, participating stakeholders advocated, should be based upon incentives rather than mandates, even as the Delphi respondents noted a need for broad measures to increase the employment and improve the socioeconomic status of persons with disabilities.

Respondents suggested that wireless technologies for persons with disabilities should be mainstream technologies with accessibility features rather than wholly separate assistive technology (AT) devices. Respondents also generally were in agreement on the major issues pertaining to access/awareness, economics, policy, and technology of wireless technologies. From the wide array of stakeholders involved, the importance placed on heightening access and awareness emerged as the leading concern. Also, a preference emerged for market-based or voluntary solutions and for mainstream universal design (UD) devices rather than an AT approach.

Discernable blocs of support for the policy options were evident. While the stakeholder groups taking part in the process agreed on a common set of desired ends, they diverged on the means to achieve them. Options receiving the greatest breadth of support related to heightening access to and awareness of wireless devices. Each of the five options received a majority of support; nevertheless, several options revealed the existence of conflicting groups. Most notably, the first access/awareness option - the development of new programs through federal and state agencies, received 78 percent of support regarding its feasibility. Supporters argued that the federal government should play a larger role in promoting awareness of and access to wireless technologies for persons with disabilities. Yet a notable 22 percent of respondents expressed reservations over the same option. Several of these stakeholders noted that both support and funds for this option were lacking in the FCC and other federal agencies.

14.5 Findings and Outcomes

The study findings strongly suggested the need for the development of an expanded array of policy approaches to address the current shortcomings of wireless technologies in impacting both society in general and specifically people with disabilities. The RERC uses the results of policy research activities to: 1) consult with stakeholders on activities of interest, 2) produce informative newsletters, 3) Generate filings before the FCC, and other pertinent agencies, and 4) contribute to actions in other related venues. Evidence of the efficacy of the RERC process can be inferred by the inclusion of RERC comments in FCC rulemakings (*e.g.* FCC 2005). Centre findings have also been published in periodic reports on the website and disseminated to members of the consultative policy network, and to other interested stakeholders.

The cross-cutting nature of the Centre's research suggests that the Centre could be enhanced by expansion of expert resources. To this end the RERC is developing a virtual network of technology policy experts, the "Collaborative Policy Network" (CPN) to collaborate on applied policy initiatives, and provide support to the Wireless RERC as well as to other RERCs conducting research in the telecommunications and information technology-related fields. The team will assist with monitoring of legal, regulatory, and policy activities at the Federal and State level, and help identify and develop appropriate policy response.

The Wireless RERC continuously monitors the policy environment by tracking legislative, regulatory, judicial, and industry activities related to access to advanced information and communications technology. Centre efforts in this area are expected to continue to keep constituents in the technology, policy, academic, and research communities informed on the developments and issues that impact the communications policy landscape, and to help promote increased access to wireless technologies for people with disabilities.

14.6 Acknowledgments

The authors wish to thank the participants of the Policy Delphi, and to acknowledge the research assistance of Alan Bakowski, Andrew Ward, Avonne Bell, Lynzee Head, Christine Bellordre, Jason Anavitarte, Andy McNeil, Adam Starr, and Lisa Griffin. The Rehabilitation Engineering Research Center for Wireless Technologies has been supported by the National Institute on Disability and Rehabilitation Research (NIDRR) of the U.S. Department of Education DOE) under grant number H133E060061, and H133E010804. The opinions contained in this paper are those of the author and do not necessarily reflect those of the DOE or NIDRR.

14.7 References

Baker PMA, Moon NW, Bakowski A (2007) Access to wireless technologies for people with disabilities: findings of a policy Delphi. Wireless RERC/Center for Advanced Communications Policy (CACP), Atlanta, GA, US

Baker PMA, Moon NW, Ward AC (2006) Virtual exclusion and telework: barriers and opportunities of Technocentric Workplace Accommodation Policy. WORK: A Journal of Prevention, Assessment and Rehabilitation, 27(4): 412–430

Baker PMA, Bellordre C (2003) Factors influencing adoption of wireless technologies – key policy issues, barriers, and opportunities for people with disabilities. Information Technology and Disabilities, 9(2). Available at: www.rit.edu/~easi/itd/itdv09n2/baker.htm (Accessed in December 2007)

Cornish E (1977) Study of the future. World Future Society, Washington, D.C., US

Dalkey N, Brown B, Cochran S (1970) Use of self-ratings to improve group estimates. Technological Forecasting, 1(3): 283–291

Linstone HA, Turoff M (eds.) (2002) The policy Delphi: techniques and applications. Reprint edition, New Jersey Institute of Technology, Newark, NJ, US. Available at: www.is.njit.edu/pubs/delphibook/index.html (Accessed in December 2007)

Wireless RERC (2003) Policy and regulatory assessment: factors influencing adoption of wireless technologies: key issues, barriers and opportunities for people with disabilities. GCATT, Georgia Institute of Technology, Atlanta, GA, US. Available at: www. wirelessrerc.org/publications/policy-briefs/ (Accessed in December 2007)

Turoff M (1970) The policy Delphi. Journal of Technological Forecasting and Social Change, 2(2): 149–172

U.S. Census Bureau (2006) P70-107: Americans with disabilities: 2002. Washington, D.C., US

Chapter 15

Gaze Interaction with Virtual On-line Communities

R. Bates, H.O. Istance and S. Vickers

15.1 Introduction

On-line real-time 'immersive' communities, such as SecondLife, are becoming increasingly popular as a means of interacting and doing things together with friends and new acquaintances. These communities represent users as avatars, through which a person may be represented by a virtual self of any shape, size, colour or other appearance, with interaction taking place in a virtual 3-dimensional world. A disabled user may construct an avatar which can reveal, or hide, their disability from other people in the virtual community, and offer a different experience from the real community.

Figure 15.1 shows the user's avatar in the foreground with the camera or viewpoint placed behind the avatar. In the distance is another avatar. Menu controls for the application are placed upper left, and camera placement and avatar movement controls are on the lower right on moveable transparent overlay panels. Frequently used commands are on buttons along the bottom of the screen. Much of the scene is animated, including trees which move in the virtual breeze, as well as running water, video advertising boards, and a tram that has just passed from view from the station on the right and from which the distant avatar has just alighted. The avatar may interact with and manipulate many of these objects, creating quite a realistic scenario.

This paper posits that while participation in these communities can offer a high-level paralysed disabled user privacy in terms of not revealing their disability during highly varied interaction with many others, this privacy will be compromised unless suitable interaction devices and techniques are available for fast communication. Eye gaze offers a high bandwidth of communication and has been very successful in enabling disabled users to access desktop applications. The same success exists potentially for 3D virtual communities if suitable ways of using gaze can be designed. The paper will show that using gaze pointing merely to emulate a mouse is not sufficient. Interaction techniques with interfaces to support these which are especially adapted to use with eye gaze are needed for the benefits of on-line virtual communities for severely disabled users to be fully realised.

Figure 15.1. 'Second Life' created by Linden Labs – www.secondlife.com

15.2 Who Do I Wish to Virtually Be?

15.2.1 Who is Me? The Appearance of Disability

Merely looking at the avatars in SecondLife, it would appear that many users regardless of disability choose to project a stylised version of themselves, rather than an actual 'close-as-possible' appearance. There are very few overweight-looking avatars, for example, in spite of the prevalence of this condition in the western adult population. A recent article in 'Disability Now' (Stevens, 2007) notes that, for some people who have a disability, SecondLife is an opportunity to escape from their 'impairment' and appear 'able-bodied' again. However, there is also a group of disabled people who wish to appear disabled as this reflects the reality of who they are. In either case, however, disclosure is optional. This raises the problem of the individual human right to privacy for disabled users, in other words: disabled users should have the choice for other users not to know about their disability when they are in a virtual world. This is just the same choice as other users have who may represent themselves as they wish.

15.2.2 Time, Disability and the Problem of Privacy

The Barrier of Control
Interaction and control of an avatar in virtual worlds is typically achieved by desktop mouse and keyboard, usually with the mouse controlling the avatar

movement and the keyboard for text input and 'chatting' in a similar manner to a typical on-line chat room. Consider a scenario of a disabled user with a high level of paralysis. What if they could not use a conventional mouse and keyboard, yet they wished to be as 'able-bodied' as any other user in the virtual world? They are now presented with a barrier of how they might control their avatar rapidly, effectively and hence efficiently so it does not 'appear' they have difficulties in computer control caused by their disability. It might seem simple to 'appear normal' as an avatar, but there is considerable subtlety required. This can be summarised: "Animated characters are judged for their level of interactivity with the user and their believability. This judgement is based not only on the characters' appearance, but moreover in their behaviour, the personality they portray, their expressions, their actions, their moods and the emotions they are able to trigger" (Romano, 2005). So the avatar controlled by a disabled user should have the option of 'being indistinguishable from the rest'. This is reminiscent of an avatar Turing (1950) test – where, when talking and interacting with an unknown avatar controlled by a disabled user, for that avatar to appear and be declared 'virtually able-bodied' and so preserve privacy and not reveal the disability of the user, one should not be able to detect any difference between the behaviour of that avatar and any other.

The Control Demands of a Virtual World

This barrier of being able to control the appearance of on-line presence is becoming more critical for disabled users with the advent of ever more feature-rich interaction within on-line communities, striving for ever more realism. This move to realism can be shown if we examine four different levels of interaction with others; email, chat rooms, basic avatars and more advanced avatars. The first two of these are well known. If an email is sent to someone who is disabled, the reply when received may give no indication of the disability of the author. The text-only chat room may also protect the privacy of the disabled user if he or she can respond with text entry in a reasonably timely manner. The basic type of avatar can have a library of gestures and actions, as well as text entry, under mouse and keyboard control, as can be found in SecondLife at present. One can envisage a situation in the near future where the users' real motions, gestures and expressions can be tracked and transferred in real time to the avatar thus increasing the level of expression. The Nintendo Wii incorporates basic motion tracking of a hand held device at present, and a recent release of SecondLife includes speech communication between avatars within a certain range. Avatars 'instrumented' in this way can be characterised as advanced avatars. This represents a hierarchy of a progressively increasing communication load on (disabled) users that threatens their choice of disability privacy (Figure 15.2).

The figure highlights the increasing difficulty disabled users will have as the demands of communication time increase (if you are slow in responding to a communication or action from another on-line person, then perhaps you have difficulties). Also the amount of data you may need to generate to enable complete communication or control over your on-line presence may reveal difficulties if you cannot generate this data quickly enough. For email a disabled user has time off-line to generate the text of the mail and any data it might contain (a low bandwidth of communication), and an email conversation is not immediate: responses are expected to take some time. For a chat room, again, typically only text is generated (a low amount of data), but this text must be delivered moderately quickly before the conversation moves on too far, giving a time constraint, a threat to privacy and a moderate exposure risk. Difficulties start to arise when we move to more 'real' representations of the user. Here with an avatar, text must be generated (or speech may be used and relayed through the avatar) in a time critical manner, and in addition the movement, posture and expression of the avatar must also be controlled convincingly, requiring a high volume of control data again in a time critical (and synchronised) manner.

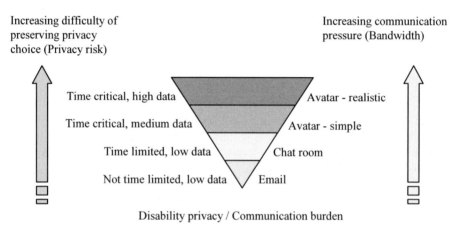

Figure 15.2. Disability privacy burden by on-line meeting type

15.3 Interaction in Second Life

To survey interaction and the burden of control in Second Life, a break down of the main types of user tasks is needed. Previous work (Hand, 1997) proposed a taxonomy of main control and manipulation areas present in virtual environments (1 to 3 below), to which communication can be added as a fourth task type:

- locomotion and camera movement - moving your avatar and moving your viewpoint on the world;
- object manipulation – creating and manipulating objects;
- application control – controlling functions of the application, menus *etc.*;
- communication – chatting, generating text.

15.3.1 Task Control Requirements

It is useful to examine briefly what actions are required for each of these task areas, assuming the use of a conventional mouse and keyboard for control.

Locomotion and Camera Movement
These both may be controlled by their respective on-screen buttons mounted in nearly transparent moveable overlays (Figure 15.1 bottom right) by simply clicking the mouse on the arrows, with continuous motion accomplished by holding down the mouse button. There are also keyboard shortcuts for avatar movement, by holding down the keyboard arrow keys, and the 'F' for flying, and also for the camera. An additional 'mouse view' mode is available where the avatar moves toward the mouse cursor at all times.

Object Manipulation
This is only accomplished via the mouse. To manipulate an existing object the mouse is clicked on that object, when a pie-menu appears with object affordances which may simply be chosen (Figure 15.1 bottom right), so for example a chair might present the choice 'sit here'. To create an object 'create' is selected and a dialog box is presented with object choices. The object type and properties are selected from the dialog box and then the object is created by clicking in the environment where the object is required. To manipulate that object a series of axes are shown and by dragging any of these the object size and orientation may be changed.

Application Control
This is mostly accomplished via the mouse, with few of the application commands and menus fully accessible via the keyboard. Menus must be opened by the mouse (Figure 15.1, top left) and although there are some keyboard shortcuts, most functions on the menu are accessible via mouse only. Changing the appearance of the avatar is a major function and is only accessible via the mouse, with few keyboard functions available apart from moving selected sliders and tabbing between buttons.

Communication
Communication is via text generation, or speech relay via a microphone. Text is typed into a chat box placed at the bottom of the screen, and text generated in a conversation is displayed and scrolled on the lower left of the screen. Speech relay allows the voice of the user to be heard within a short distance of their avatar, together any nearby avatars to be heard by them. The enabling of speech results in far fewer interaction problems for communication for most disabled users by removing typing from communication. However, not all disabled users may wish to use or be capable of using speech.

Control Summary

The control requirements of Second Life may be summarised by control source and task domain (Table 15.1). This summary allows us to determine what combinations of control are required to interact with Second Life. For example, a mouse and speech enables full interaction, as would a mouse and keyboard, but a mouse or keyboard alone would not enable interaction. These findings will now be used to determine how a disabled user might interact with Second Life using alternative interaction techniques.

Table 15.1. Control requirements for task domains

Task domain	Control source		
	Mouse	**Keyboard**	**Speech**
Locomotion and camera movement	✓	✓	✗
Object manipulation	✓	✗	✗
Application control	✓	Partial	✗
Communication	✗	✓	✓

Key: ✓ full control may be achieved, ✗ no control possible

15.3.2 Using Gaze for Interaction

Users with severe disability can not always operate a standard mouse or keyboard; and probably cannot move their head well; they may have some speech but may have problems in being recognised or problems due to aided respiration; but they would retain eye movement, since eye control is retained in all but the most advanced cases of ALS.

Gaze tracking has been shown to be an effective means of computer control for users with high level paralysis (Bates, 2002; Bates and Istance, 2003, 2004) and has been used effectively in eye controlled games (for example, Isokoski and Martin, 2006) and in immersive environments (Cournia *et al.*, 2003; Tanriverdi and Jacob, 2000); and finally, on-screen keyboards are readily available with basic versions included with the accessibility tools of Windows. One approach to using eye-gaze is mouse emulation by gaze tracking (placing the mouse cursor where the user is gazing on the screen), and keyboard emulation via an on-screen keyboard. The underlying application is unaware that the pointer movement, button and keystroke events originate from a gaze-based device. However, simply using the eye as a mouse has its own problems.

Istance *et al.* (1996) presents a set of issues associated with using the eye to control desktop applications. Some of these will be apparent in controlling and interacting with environments such as SecondLife. One of these was the need for switching the gaze point between the point of input and the point of feedback. If gaze selection is active then the input control looses focus during this action, and the point of feedback may gain focus.

It is important to make an initial assessment of whether mouse emulation by eye-gaze is a satisfactory solution to meet the needs of disabled users when interacting with SecondLife and similar environments. To enable high-level paralysis disabled users to use this virtual world, an assessment was required to determine if this alternative means of interaction would enable successful, rapid, effective and hence efficient interaction with SecondLife and so give users with a disability full control over their avatar in the virtual world whilst not revealing their disability.

15.4 An Experiment with Gaze Interaction in a Virtual World

An investigation was conducted to determine the possibility of interaction via gaze with SecondLife. In this initial evaluation work two expert users who were highly experienced in using gaze-based control, and also experienced and familiar with SecondLife were chosen to attempt gaze control with the on-line environment. The issue here was not to conduct large trials of SecondLife, but to simply assess if mouse emulation by gaze control was practical and effective. Both gaze control and a baseline of a normal desktop hand mouse were tested. Text entry was via the Windows on-screen keyboard for both mouse and gaze to give equivalence in text generation task results. Five tasks were constructed from the essential task domains (Table 15.1), in which an avatar was required to perform a short set of actions under the control of gaze (tracked by an SMI REDII remote infrared eye tracker). The evaluators were sat at 60cm from a 17" monitor, with the eye tracker giving an approximate accuracy of ±0.5 to 1cm in cursor position on the screen at this distance. The tasks can be summarised as follows:

- locomotion – the subjects were required to make the avatar walk along a path approximately two paces wide that follows a rectangle 50 by 25 paces in size around a park, negotiating past trees and other distracting obstacles;
- camera movement – the camera was to be moved from behind the avatar around clockwise to face the avatar, and then moved overhead to view the avatar from above, and then moved away to high above the avatar so the complete park could be seen, and finally placed back in its original position by retracing the path;
- object manipulation – a cube was to be created, resized to as close to 2m cube as possible;
- application control – the avatar was to be changed in appearance by changing the hair colour to blonde;
- communication – the avatar was to chat with another avatar, generating the following "The weather here is nice, it is always sunny and warm".

The overall completion times for each task were recorded, together with a count of errors occurring during the task (Table 15.2). Finally the evaluators were asked to make comments on gaze controlled task areas they found either easy or difficult in comparison to the baseline desktop mouse.

Table 15.2. Control performance for task domains

| Task domain | Control Source / Task time (s) and Error counts | |
	Mouse	Gaze
Locomotion	48s (3 path)	88s (2 path, 2 distraction)
Camera movement	50s	122s (2 accuracy, 8 feedback)
Object manipulation	35s	71s (3 accuracy)
Application control	20s	194s (4 accuracy)
Communication	60s (=11 wpm)	224s (8 acc., 10 feedback)
		(=3 wpm)

Four main types of errors within the tasks in the gaze condition were identified and were defined as follows:

- path deviation – movement or wandering from the chosen or desired path made by poor positional control of the avatar direction;
- distraction – errors particular to gaze control where the gaze is distracted to another object in the world and since gaze is controlling motion direction; that motion is also pulled toward the distraction;
- accuracy – simple pointing accuracy problems due to the inaccuracy of gaze tracking and pointing resulting in difficulties placing the cursor on small controls;
- feedback – the continued gazing between the interaction point where gaze is manipulating a control, and the location of the actions or feedback caused by manipulating that control.

We can now discuss the results and examine the effectiveness of gaze control against the baseline of the desktop mouse for each of the task domains.

15.4.1 The Potential for Gaze Control by Task Domain

Locomotion
The task times for gaze control were longer than for the mouse, though not excessively so, and path deviations were also similar, showing that gaze control of locomotion was effective, with the evaluators simply looking where they wished to go and the avatar following that gaze point. The major issue was of distraction,

where the gaze of the user was pulled away, even temporarily, from the desired destination to some other in-world object, hence moving the avatar toward that object involuntarily. This is illustrated in a sequence (Figure 15.3, left to right) where the gaze is distracted by a tree, resulting in the avatar walking into the tree instead of staying on the path.

Figure 15.3. Locomotion and distraction

Examining the comments made by the evaluators, gaze driven locomotion was regarded as viable: "Steering by eye worked well, all I needed to do was to look at the object I wished to walk towards", "I feel that this could be a very rapid means of steering, just look and you go right there, it could be better than a mouse". However, the main issue of distraction was noted "You can't look around while you are walking without walking off the path – it needs some form of clutch to decide if you are steering by eye, or just looking around". This problem could be overcome by introducing some disambiguation between gaze controlled locomotion (pointing at where you wish to go) and gaze 'looking around' in the world. Possibly by adding an in-world object 'sticky' gaze point (the user gazes at where they wish to go and indicates this destination) gaze is then freed from locomotion control and the user is free to look around until they gaze back at the 'sticky' gaze point and regain control of gaze-driven locomotion.

Camera Movement
Here task times were considerably longer for gaze than mouse. There were accuracy errors due to the small size of the camera movement control and the inaccuracy of gaze pointing, and also a large number of feedback errors caused by the difference in screen location between the camera control and the view of the avatar on the screen. This is illustrated in a sequence (Figure 15.4, left to right) with continual gazes back and forth between the camera control and the avatar (to determine the new camera position) resulting in the camera tracking and responding to the movement of the cursor – note how the camera orientation rapidly moves between the user looking at the control (left and right images), and the user looking at the avatar (centre image). This separation of control point from point of feedback could be overcome by placing the control at the point of feedback, possibly by making the control a 'toolglass' – a translucent control that allows viewing of the feedback in the world at the point of control.

Figure 15.4. Camera movement and feedback

Object Manipulation

Task times were longer and error rates higher for gaze than the mouse, but not excessively, with errors caused by difficulty in placing the gaze cursor on the small handles on the objects. However, the evaluators commented on how easy to use the pie-menu was for gaze (shown in Figure 15.1) as it was both large (easy to acquire with gaze) and also placed at the point of interaction so there were no feedback problems (it appears wherever the cursor is in the world, and is translucent so as not to obscure the world behind it).

Figure 15.5. Object sizing by gaze

It was of particular interest that when once the handles of the object were acquired, the object could be resized very effectively through the appearance of rulers extending from the object, with the user simply needing to gaze at the required measurement (thus placing the cursor on that measurement) for the object to be correctly resized. This is illustrated where the user is gazing at 4m on the ruler thus resizing the object to 4m (Figure 15.5). However, note that feedback of the object dimensions appears at the top of the screen – if the user gazed at these figures while controlling the object, the object sizing would follow the gaze and be distorted.

Application Control
Here the task times were considerably longer and error rates higher for gaze due to the limited pointing accuracy of gaze control and the small buttons on the application control interface. In this task gaze is simply emulating the mouse, with no coupling of gaze position to control and feedback (as in locomotion and camera movement) and with interaction and feedback in essentially the same screen area. The gaze inaccuracies here can be overcome in the same way as has been shown on a 2D interface by employing a magnification tool (Bates *et al.*, 2002) that would temporarily magnify the interface (zoom in) to allow easier manipulation of small controls.

Communication
The use of an on-screen virtual keyboard was slow for the mouse (11 words per minute) and very slow for gaze (three words per minute), thus presenting a significant communication problem for our disabled user. Compared with manual typing speed of about 40 wpm (Majaranta and Räihä, 2002) this is very slow. In addition, there were considerable accuracy problems giving a long task time, and many repeated gazes at the text as it was being generated to see if errors had been made. This is illustrated by red lines (gaze paths) and red dots (gaze fixations) showing the many gazes between keyboard and chat box on SecondLife when writing only 10 characters (Figure 15.6).

Figure 15.6. Feedback while gaze typing

This continual feedback gazing could be reduced by a second modality such as speech reading aloud the characters as they are being generated, thus relieving gaze from monitoring feedback and leaving gaze to solely generate input. Another approach would be to incorporate a third party dedicated text generation application into the world, such as Dasher, a pseudo 3-dimensional typing tool which generates text via gaze directed 'flying' into letters, that might lend itself well to 3D virtual worlds. With this tool more than 25 wpm have been achieved (Ward and MacKay, 2002). Another approach is sentence expansion by possibly using a pie-menu (such as that in Figure 15.1) with pre-defined sentences available for selection. However, communication speed will remain a problem for many disabled users.

15.5 Conclusions and the Future

15.5.1 Discussion

The data presented in this paper form the basis for preliminary observations and are intended to indicate directions for research rather than to provide definitive answers. However, it would appear that simply treating eye gaze as a way of emulating mouse behaviour will not deliver the speed of interaction necessary to safeguard the privacy choices of disabled users.

The types of errors found suggest a need for a lightweight clutch mechanism whereby gaze control can be activated and deactivated quickly and effortlessly. A control action can be applied by eye and then gaze control is deactivated to enable the user to see the effect of the action and to look at the objects nearby. Gaze-based gestures are not widely used and represent a potentially fast way of achieving this. This mechanism will be successful only if the effort involved in using it is considered to be justified by the benefits it affords. It brings the additional issue of mode feedback. Is eye control currently active or not, and what is the best way of indicating this? The 'sticky' gaze point referred to earlier represents another modal gaze specific interaction technique, where careful study of workload and feedback will be needed to ensure sufficient usability.

The question of the separation of the point of input control action and feedback position is still an issue and has been partially addressed in SecondLife with the provision of moveable transparent panels of icons, and pie-menus that appear over objects in the world. Controlling when and where these appear by gaze, as well as extending the functionality of these is likely to be a very fruitful area of work to make gaze a successful modality for the disabled user. 'Toolglass' technology where the action represented by the icon on the toolglass is applied to the object in the world directly behind the icon represents a very useful direction to examine in identifying candidate gaze interaction techniques (Kurtenbach et al., 1997).

15.5.2 Conclusions

Using gaze to interact with virtual environments as a modality for users is still at an early stage of investigation and holds much promise when used with other modalities. The challenges of using gaze alone to interact in real-time (or close to real-time) with virtual environments are considerable, but if these can be met then there will be greater opportunities for disabled users to participate fully in virtual communities. Until this control is fully realised disabled users may feel that they have challenges to their choice of on-line privacy.

15.6 References

Bates R (2002) Have patience with your eye mouse! Eye-gaze interaction with computers can work. In: Proceedings of the 1st Cambridge Workshop on Universal Access and Assistive Technology (CWUAAT'02), Trinity Hall, University of Cambridge, Cambridge, UK

Bates R, Istance HO (2003) Why are eye mice unpopular? A detailed comparison of head and eye controlled assistive technology pointing devices. Universal Access in the Information Society, 2(3): 280–290

Bates R, Istance HO (2004) Towards eye based virtual environment interaction for users with high-level motor disabilities. In: Proceedings of the International Conference on Disability, Virtual Reality and Associated Technologies (ICDVRAT 2004), New College, Oxford, UK

Cournia N, Smith JD, Duchowski AT (2003) Gaze vs. hand-based pointing in virtual environments. In: Proceedings of the International ACM CHI 2003 Conference on Human Factors in Computing Systems (CHI 2003), Ft. Lauderdale, FL, US

Hand C (1997) A survey of 3D interaction techniques. Computer Graphics Forum, 16(5): 269–281

Isokoski P, Martin B (2006) Eye tracker input in first person shooter games. In: Proceedings of the 2nd Conference on Communication by Gaze Interaction: Communication by Gaze Interaction (COGAIN 2006), Turin, Italy

Istance HO, Spinner C, Howarth PA (1996) Providing motor-impaired users with access to standard graphical user interfaces (GUI) software via eye-based interaction. In: Proceedings of the 1st European Conference on Disability, Virtual Reality and Associated Technologies (ECDVRAT'96), Maidenhead, UK

Kurtenbach G, Fitzmaurice G, Baudel T, Buxton B (1997) The design of a GUI paradigm based on tablets, two-hands, and transparency. In: Proceedings of the ACM CHI 97 Conference on Human Factors in Computing Systems (CHI 97), Atlanta, GA, US

Majaranta P, Räihä K-J (2002) Twenty years of eye typing: systems and design issues. In: Proceedings of Eye Tracking Research and Applications (ETRA 2002), New Orleans, LA, US

Romano DM (2005) Synthetic social interactions. In: Proceedings of the H-ACI Human-Animated Characters Interaction Workshop, British HCI 2005, the 19th British HCI Group Annual Conference, Napier University, Edinburgh, UK

Stevens S (2007) Virtually the same. Disability Now, May 2007. Available at: http://archive.disabilitynow.org.uk/search/z07_05_My/virtually.shtml (Accessed on 7 December 2007)

Tanriverdi V, Jacob RJK (2000) Interacting with eye movements in virtual environments. In: Proceedings of the International ACM CHI 2000 Conference on Human Factors in Computing Systems (CHI 2000), The Hague, The Netherlands

Turing A (1950) Computing machinery and intelligence. Mind, LIX(236): 433–460

Ward DJ, MacKay DJC (2002) Fast hands-free writing by gaze direction. Nature, 418(6900): 838

Chapter 16

The Resolution Race: Perpetuating Inaccessible Computing

G. Hughes and P. Robinson

16.1 Introduction

While an average computer user clamours for the increased screen real estate gained by increasing a monitor's resolution, users with visual disabilities struggle as the information gets smaller. Unfortunately for the users with visual disabilities, higher resolutions have become increasingly common over the past few years. Due to this increased average resolution, software and webpage developers often neglect users of lower resolutions in favour of using all of the available space on high-resolution displays. This paper aims to investigate the needs of users with visual disabilities and how they are negatively affected by developers who design interfaces which can not be used efficiently at multiple resolutions. Possible solutions to this problem will also be discussed.

16.2 The Resolution Race

Just as chip developers have striven to cram more transistors into a set amount of space, display developers have aimed to squeeze more pixels into each square inch of a display. A monitor's resolution is defined as the number of columns and rows of pixels that can be displayed. A monitor with a high resolution, such as 1600x1200, displays more pixels than a monitor with a lower resolution, such as 800x600. This increase in pixels results in an increased amount of space to display information because current graphical user interfaces render information with a set pixel height and width regardless of the monitor's resolution. For example, the Save icon in Microsoft Word 2003 measures 14 pixels square. On a 21 inch, 4:3 monitor with a resolution of 800x600 this icon will measure 7.47 mm square. On the same monitor with a resolution of 1280x1024 the icon will measure less than 4.68 mm square. Similarly, most text rendered by current operating systems would also undergo a decrease in size by a factor of about 1.6. Decreasing the size of fonts and icons by increasing the output resolution maintains all of the information while making more room to display additional icons, text, menus, or other visual information. As long as the text is still legible, this is an obvious benefit for the average user, but it is apparent that this size decrease may cause problems for individuals with visual disabilities. Aside from those users with visual disabilities, higher resolution also causes problems for individuals with motor impairments because targets become smaller requiring finer mouse movements.

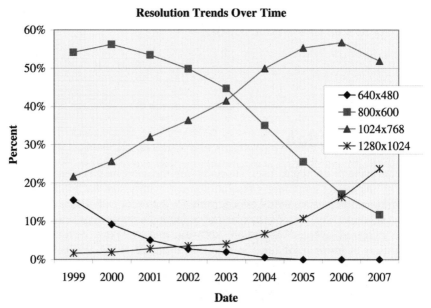

Figure 16.1. Graph of monitor resolutions used by more than 10% of users (TheCounter. com, 2007)

16.3 Human Vision

To better understand the difficulties that individuals with disabilities may encounter with higher resolutions, one must first understand the basics of human vision and what it means to have a reduced visual acuity. The term "visual acuity" refers to the ability of our visual system to distinguish fine details. The smallest detail that can be perceived by the human visual system is measured in arc seconds by optometrists using the Snellen metric. The average human can perceive around 1 arc minute of detail, measurable with the Snellen eye chart. In the 19th century, Hermen Snellen maintained that if a human eye could distinguish details down to 1 arc minute in thickness, then a human could read a letter where the strokes subtended 1 arc minute and the distance between strokes subtended 1 arc minute (Thomson, 2005). For example, the letter E would be drawn so that it would be 5 arc minutes tall and 4 arc minutes wide (see Figure 16.2). The Snellen chart is viewed at a standard distance of 20 feet in the US, and 6 metres in the UK. We can determine the width of one arc minute as follows:

$$1' = \left(\frac{2\pi}{21600}\right) rad$$

$$\tan\left(\frac{2\pi}{21600}\right) = \left(\frac{x}{6m}\right)$$

$$= 6000mm \times \tan\left(\frac{2\pi}{21600}\right)$$

$$x \approx 1.745mm$$

The entire letter E would subtend 5 arc minutes, therefore it would measure 8.725 mm in height. Using the Snellen chart, visual acuity is represented as a fraction, with the chart viewing-distance as the numerator. The denominator represents the distance at which a particular letter would subtend 5 minutes of an arc for the average person. For example, if a patient's visual acuity were measured at 6/9, the smallest letter they would be able to read at 6 metres would subtend 5 arc minutes at 9 metres for the average human; the average human would be able to read the same line from 9 metres that the patient was only able to read from 6 metres.

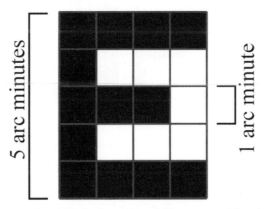

Figure 16.2. The manner in which the letter E would be composed for viewing on a Snellen Eye Chart

In the US the term "legally blind" is used to describe someone with vision that cannot be corrected to better than 20/200 (6/60 in the UK) in the better eye, or having a field of vision less than or equal to 20 degrees in the better eye (Social Security Administration, 2005). This classification is similar to the UK Department of Health classification of Registered Partially Sighted (Department of Health, 2003). If a legally blind patient were to look at a Snellen chart, the smallest letters that would be visible from 6 metres would be visible for someone with average vision at 60 metres. Therefore, the size for the letter E on the 6/60 line would be 10 times bigger, or 87.25 mm. This is not a theoretical assumption, but rather how visual acuity is defined.

Using this understanding of human vision, we can determine the smallest possible details visible on a computer monitor to a user. This approach disregards colour or contrast information and merely focuses on the smallest distinguishable detail under ideal circumstances. Many characters are composed of a single line of pixels (See Figure 16.3). If the pixels are outside of visual perception under ideal conditions, then regardless of their colour or contrast the letters would be indeterminable. In practice, the smallest visible detail may be quite a bit larger and will vary depending on the contrast and colours of the information along with other factors such as the lighting conditions.

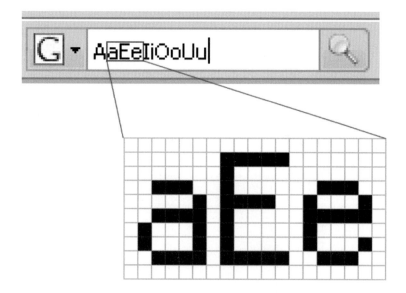

Figure 16.3. Each character in the Firefox search bar is composed of single lines of pixels

An average computer user will be about 30 to 60 cm away from a computer monitor. From a distance of 60 cm a user with average vision will be able to see details that are as small as 174 μm square. From Table 1.1 we see that a 4:3, 21 inch monitor with a resolution of 1280x960 displays pixels that measure 333 μm square. This pixel size is clearly within the visual range of the average human since who will be able to perceive detail as small as 0.42 pixels. However, from 60 cm away the smallest detail that an individual with 6/60 vision can perceive measures 1.74 mm. In order to be able to perceive all of the details displayed on the same monitor this user would need to be at most 10.7 cm away from the monitor. It would be impossible for an individual to simultaneously be 10.7 cm or less from every point on a 21 inch monitor, and thus the user would need to move around to see all of the information. Additionally, it can be seen that the Snellen metric scales linearly. Therefore in order to read information with the same ease as an individual with average vision who sits 60cm away from a monitor, someone with 6/60 vision would need to be 6 cm away from the monitor.

Table 16.1. Shows smallest detail that an individual can see on a monitor from 60cm away

Monitor Size (inches)	Resolution	Pixel size (μm^2)	Smallest visible detail (in pixels)		
			6/6 vision	6/18 vision	6/60 vision
21	640x480	666.75	0.26	0.79	2.62
21	800x600	533.40	0.33	0.98	3.27
21	1024x768	416.72	0.42	1.26	4.19
21	1152x864	370.42	0.47	1.41	4.71
21	1280x960	333.38	0.52	1.57	5.24
21	1400x1050	304.80	0.57	1.72	5.73
21	1600x1200	266.70	0.65	1.96	6.54
21	1900x1200	238.06	0.73	2.20	7.33
30	2560x1600	252.41	0.69	2.07	6.91
22.2	3840x2400	124.52	1.40	4.20	14.02

If the same visually impaired user decreased the monitor's resolution to 800x600, then the user is able to sit about 18 cm away from the monitor. This almost doubles the distance between the user and the monitor and thus is a preference for many users with visual disabilities. When using a laptop in particular, this extra 8cm can make a huge difference to a user with a visual disability. Decreasing the resolution to 640x480 would make the information even larger; however this is literally no longer an option on current hardware and versions of Microsoft Windows.

A user with a visual impairment may need to get even closer to a monitor than these figures suggest because of the colour or contrast of the presented information. Human vision is far more complex than visual acuity alone; colour, contrast, depth, and field of view are also important factors. However, when solely examining resolution, visual acuity is the only applicable metric. This mathematical analysis does not finitely determine the size that information should be, but rather the smallest detail that can be perceived under ideal conditions, which generally don't exist in the real world. Therefore, information would need to be even larger for an individual with a visual impairment.

16.4 Resolution-dependence Problem

Although using low resolution is an efficient computer adaptation method for users with visual disabilities it is becoming a less viable option due to the resolution race. As display manufactures are racing to create higher resolution displays and video cards, developers are developing software exclusively for these higher resolution displays. This forces users with visual impairments to switch to less efficient adaptation methods such as using screen readers

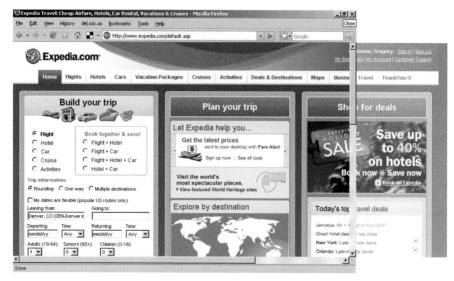

Figure 16.4. Expedia.com homepage, designed for resolutions greater than 800x600

Figure 16.5. cnn.com homepage, designed for resolutions greater than 800x600

Web developers are beginning to design sites for use on displays with a resolution of at least 1024x768. Although these sites will still be displayed on a lower resolution display, they will contain a horizontal scrollbar. A horizontal scrollbar is a user-interface element shunned in the world of web design because it is terribly inefficient. Amazon, Expedia, Yahoo Finance, CNN, and The US Whitehouse are just a few examples of web pages that contain a horizontal scrollbar when displayed at resolutions lower then 1024x768. It is widely known that the use of a horizontal scrollbar is inefficient and indicative of a poor user interface because it inhibits the user. Older users often do not even notice the presence of a horizontal scrollbar and thus miss any content outside of the main window (Chadwick-Dias *et al.*, 2003). Therefore, one must wonder why web designers are neglecting this relatively large group of web users. More importantly, web pages created by US government organizations must be made accessible to users with disabilities under the Americans with Disabilities Act (ADA). It is clear that web pages designed specifically for higher resolution are more difficult to navigate for users with disabilities when compared to resolution-independent web pages. Therefore, since it is very simple to design a resolution-independent webpage, is it illegal under the ADA for government sites to be designed for higher resolutions?

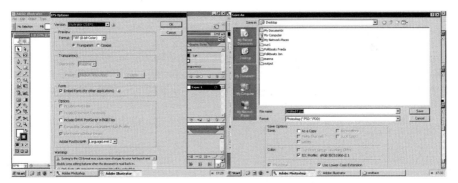

Figure 16.6. Shows the Adobe Illustrator CS2 EPS options dialog (left) where some warnings are not visible. Options are also missing from the Photoshop CS2 save dialog (right) when using a resolution of 800x600.

Like web developers, software developers also design interfaces on the assumption that the majority of users will be using a high-resolution display. Unfortunately, the consequences of resolution-dependent software can be more severe than resolution-dependent web pages. This is because software interfaces do not always include a horizontal scroll bar if the resolution is too low. For example, Adobe's Photoshop CS2 is designed to be used with a resolution of at least 1024x768. Some features of this software are inaccessible to a user with a resolution of 800x600 (See Figure 16.5). Many pieces of software require that the display resolution be set to at least 1024x768; examples include the following: Adobe's Photoshop CS3 (Adobe, 2007a), Adobe's Premier Pro CS3 (Adobe, 2007b), Autodesk's AutoCAD 2008 (Autodesk, 2007), Digidesign's Pro Tools LE 7.3 (Digidesign, 2007), IBM's WebSphere Business Modeler Advanced (IBM, 2007), Microsoft's Visio 2007 (Microsoft, 2007a), Microsoft's Works 9 (Microsoft, 2007b) and National Instruments' LabVIEW 8.5 (National Instruments, 2007). These examples alone are hardly catastrophic, but it is indicative of a growing problem for individuals with visual impairments who rely on low resolutions. Just as 640x480 is no longer supported, 800x600 will be replaced by 1024x768, causing many problems for individuals with visual disabilities.

16.5 Resolution-independent Interfaces

Many sources emphasize the importance of developing accessible software and websites for users with disabilities. These sources generally focus on the needs of users who are blind and deaf, including technical recommendations such as filling in the ALT image tags and efficiently using CSS. However, they often overlook, or underestimate, the importance of developing web pages which are resolution independent as a means of adapting for users with visual disabilities. Designing resolution-independent web pages is a trivial task; it is a conscious decision to create a web page that is resolution dependent. Basic HTML will stretch and shrink according to the page content.

Developing resolution-independent software and operating systems is *not* a trivial task. Apple Computer has filed a patent (Apple Computer, 2006) describing a concept for a resolution-independent interface. Elements, such as the previously discussed Microsoft Works icon, are stored as rendering instructions which can be effectively displayed at any resolution. This approach to user-interface graphics is similar to the manner in which vector graphics can be rendered at any resolution. Since this technology has not been released it is not yet known if it will aid users with visual disabilities.

A truly resolution-independent interface should display efficiently at a resolution of 320x240 and at 1600x1200. No such interface has yet been developed. Interestingly, such an interface would work effectively on both a 21 inch monitor and a 3 inch PDA screen. As monitor resolutions continue to increase it is becoming more and more important to develop a resolution-independent interface for use by individuals with visual disabilities.

16.6 Conclusion

It is clear that the increased use of higher resolutions is causing problems for individuals with visual disabilities. Although it is unclear how widespread the issue is, it highlights the need for increased awareness regarding the development of software which can be used by individuals with disabilities. It is critical that developers not only think about how software can be use with adaptive solutions such as screen readers, but also how their software will display at lower resolutions.

16.7 References

Adobe (2007a) Adobe Photoshop CS3: system requirements and languages. Available at: www.adobe.com/products/photoshop/photoshop/ systemreqs/ (Accessed on 21 October 2007)

Adobe (2007b) Adobe Premiere Pro CS3: system requirements. Available at: www.adobe.com/products/premiere/systemreqs/ (Accessed on 21 October 2007)

Apple Computer (2006) Resolution independent user interface design. US Patent Appliction Number 20060284878. Filed on 21 July 2006

Autodesk (2007) AutoCAD: features and specification. Available at: http://usa.autodesk.com/adsk/servlet/index?siteID=123112&id=8446045 (Accessed on 21 October 2007)

Chadwick-Dias A, McNulty M, Tullis T (2003) Web usability and age: how design changes can improve performance. In: Proceedings of the 2003 Conference on Universal Usability (CUU'03), Vancouver, British Columbia, Canada

Department of Health (2003) Form CVI: explanatory notes for consultants ophthalmologists and hospital eye clinic staff. Available at: www. dh.gov.uk/en/Publicationsandstatistics/ Publications/ PublicationsPolicyAndGuidance/DH_4083552 (Accessed on 21 August 2007)

Digidesign (2007) Support and downloads: Pro Tools LE 7.3 – FireWire Windows laptop systems. Available at: www.digidesign.com/index.cfm?navid=3&langid=100&categoryid =35&itemid=24353 (Accessed on 21 October 2007)

IBM (2007) IBM – WebSphere business modeler advanced: system requirements. Available at: www-306.ibm.com/software/integration/wbimodeler/advanced/sysreq/index.html (Accessed 21 Oct 2007)

Microsoft (2007a) Microsoft Office Visio 2007 system requirements: Microsoft Office Visio 2007. Available at: http://office.microsoft.com/en-gb/visio/HA101945411033.aspx (Accessed on 21 October 2007)

Microsoft (2007b) Minimum system requirements for Works 9. Available at: http://support.microsoft.com/kb/939451 (Accessed on 21 October 2007)

National Instruments (2007) System requirements for LabVIEW development system and LabVIEW modules: products and services: national instruments. Available at: www.ni.com/labview/requirements.htm (Accessed on 21 October 2007)

Security Administration (2005) If you are blind or have low vision. Available at: www.socialsecurity.gov/pubs/10052.html (Accessed on 5 May 2005)

TheCounter.com (2007) Available at: www.thecounter.com/stats/ (Accessed on 20 August 2007)

Thomson D (2005) Va testing in optometric practice part 1: the Snellen chart. Optometry Today, (April): 56–58

Part IV

Assistive Technology

Chapter 17

A Case Study of Simulating HCI for Special Needs

P. Biswas and P. Robinson

17.1 Introduction

Usability evaluation is an important step for any successful product design. There are different usability evaluation techniques like heuristic evaluation, cognitive walkthrough, review based evaluation *etc.*, but for assistive technology, log file based analysis is often used (Lesher *et al.*, 2000; O'Neill *et al.*, 2002). An example of a different approach, Rizzo *et al.* (1997) evaluated the AVANTI project (Stephanidis *et al.*, 1998), by combining cognitive walkthrough and Norman's seven-stage model. From the viewpoint of assistive technology, it is often difficult to find participants with specific disabilities. Petrie *et al.* (2006) use remote evaluation but still need to find disabled participants. This paper uses a different approach to evaluate assistive interfaces and presents a simulator that can predict time and possible interaction patterns for motor-impaired people undertaking tasks. The simulator is used to compare existing interfaces and evaluate new alternatives.

Our simulator takes a task definition and locations of different objects in an interface as input, predicts the cursor trace and probable eye movements in screen and task completion time, for different input device configurations (*e.g.* mouse or single switch scanning) and undertaken by persons with different skill levels and physical disabilities. The input and output of the simulator are shown in Figure 17.1. and its architecture is shown in Figure 17.2. It consists of the following three components:

The Application model models the task currently undertaken by the user. In this model we break up a complex task into a set of simple atomic tasks.

The Interface Model: Our simulator works for both able-bodied users and those with disabilities. Several types of disability impede use of a conventional mouse, keyboard and screen to interact with a computer. People with severe motor-impairment often access computers by one or two switches instead of keyboard and mouse. Visually impaired users often need to use a screen reader. The interface model decides interface types to be used by a particular user and sets parameters for an interface.

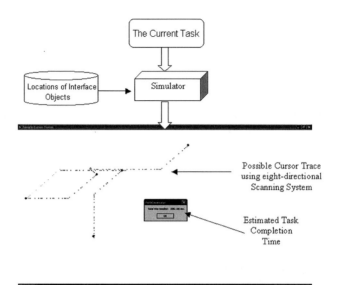

Figure 17.1. Input and output of the simulator

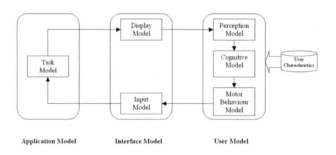

Figure 17.2. Architecture of the simulator

The User Model simulates the interaction patterns of users whilst undertaking a task analysed by the task model under the configuration set by the interface model. There is not much reported work on systematic modelling of assistive interfaces. McMillan (1992) felt the need to use HCI models to unify different research streams in assistive technology, but his work aims to model the system rather than the user. The AVANTI project (Stephanidis *et al.,* 1998) models an assistive interface for a web browser based on some static and dynamic characteristics of users. The interface is initialised according to some static characteristics (*e.g.* age, expertise, *etc.*) of the user. During interaction, it adapts itself according to some dynamic characteristics (*e.g.* idle time, error rate, *etc.*) of the user. This model does not address the basic perceptual, cognitive and motor behaviour of users and so it is hard to generalize to other applications. Our user model (Biswas *et al.,* 2005) takes a more generalized approach than the AVANTI project. It breaks down the task of user modelling in several steps that include clustering users according to their physical and cognitive abilities, customizing interfaces to suit user characteristics and logging user interactions to update the model itself. However, the objective of this model is to design adaptable interfaces and not to simulate users' performance. Keates *et al.* (2000) measured the difference between able-bodied and motor-impaired users with respect to the Model Human Processor (MHP) and motor-impaired users were found to have a greater motor action time than their able-bodied counterparts.

Our user model also uses the sequence of phases defined by Model Human Processor. It consists of perception, cognitive and motor-behaviour models. The perception model simulates the visual perception of interface objects. The cognitive model takes the output of the perception model and decides an action to accomplish the current task. The motor behaviour model then predicts the completion time and possible interaction patterns for performing that action.

17.1.1 Perception Model

Among existing cognitive architectures, EPIC (Kieras and Meyer, 1997) and ACT-R/PM (Byrne, 2001) have distinct perception models. Currently our perception model considers only vision. It takes a list of keyboard and mouse events and a sequence of bitmap images of an interface as input and produces a sequence of eye-movements and the visual search time as output.

We perceive something on a computer screen by focusing attention on a portion of the screen and then searching for the desired object within that area. The perception model models the focusing area of the user by defining a focus-rectangle within a certain portion of the screen. The focus rectangle can be moved throughout the screen either systematically (left to right, top to bottom) or randomly. The user's attention remains confined within the focus-rectangle. Recent theories of attention incorporate both top down and bottom up mechanisms of attention (Luck *et al.*, 1997; Reynolds and Desimone, 1999). The bottom-up theories are supported by implementing the colour histogram matching algorithm to find out the probable points of attention fixation and the top down theory is modelled by heuristics that govern the visual scanning process. The model can also be used to simulate effects of different visual impairments by setting its input parameters.

17.1.2 Cognitive Model

Models like KLM, GOMS (John and Kieras, 1996) *etc.* can predict interaction time for expert users, but are not an appropriate interaction model for novice users. We model the behaviour of expert users and the sub-optimal behaviour of novice users separately. The CPM-GOMS model (John and Kieras, 1996) is used to model the expert users. The sub-optimal behaviour of novice users has more randomness than this optimal behaviour. Existing cognitive architectures can be used to model novice users, but it has also been found that use of a high fidelity model (like EPIC) does not always give better results than a low fidelity model (like GLEAN)

(Kieras, 2005). In this work, we have used a probabilistic rule based system to simulate the sub-optimal behaviour.

17.1.3 Motor Behaviour Model

The motor behaviour model simulates movement limits and capabilities of users for different input devices and interaction techniques (MacKenzie, 2003). For able-bodied users, most of the motor-behaviour models are based on Fitts' Law (Fitts, 1954) and its variations (MacKenzie, 2003). For disabled users, there is growing evidence that their interaction patterns are significantly different from those of their able-bodied counterparts (Keates *et al.,* 2000; Keates and Trewin, 2005). However, we do not have any model like Fitts' law that can predict their interaction patterns. We are developing separate motor-behaviour models for different input devices based on statistical analysis of cursor traces of disabled users.

17.2 A Case Study

Many physically challenged users cannot interact with a computer through a conventional keyboard and mouse. For example, Spasticity, Amyotrophic Lateral Sclerosis and Cerebral Palsy confine movement to a very small part of the body. People with these disorders may interact with a computer through one or two switches with the help of a scanning mechanism. Two types of scanning mechanisms are commonly used for navigating through a graphical user interface. Cartesian scanning moves the cursor progressively in a direction parallel to the edges of the screen, and polar scanning selects a direction and then moves along a fixed bearing. A particular type of polar scanning (referred to as eight-directional scanning) that allows movement only in eight directions is commonly used (O'Neill *et al.,* 2000). There is little previous work on user models for scanning interfaces that can predict task completion time or interaction patterns. We have developed and evaluated a cognitive model for an eight-directional scanning system. We have assumed that our intended users have no cognitive impairment and so their cognitive model will be same as that of able-bodied users. So we validated the cognitive model through an experiment with able-bodied users.

17.2.1 Eight-directional Scanning

In this scanning technique the pointer icon is changed at a particular time interval to show one of eight directions (Up, Up-Left, Left, Left-Down, Down, Down-Right, Right, Right-Up). The user can choose a direction by pressing the switch when the pointer icon shows the required direction. When the pointer reaches the desired point in the screen, the user has to give another key press to stop the pointer movement and make a click. A state chart diagram of the scanning system is shown in Figure 17.3. A demonstration of the scanning system can be downloaded from www.cl.cam.ac.uk/~pb400/EightD.wmv.

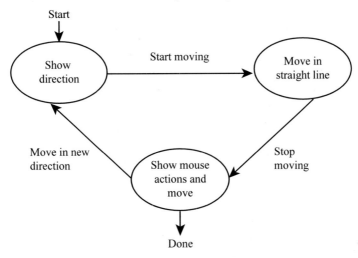

Figure 17.3. State Transition Diagram of the eight-directional scanning mechanism with a single switch

17.2.2 A Cognitive Model for Eight-directional Scanning

As mentioned previously, we have developed the cognitive model by separately modelling the optimal (expert) and sub-optimal (non-expert) behaviour of users.

Modelling Optimal Behaviour
The task of clicking on a target in an eight-directional scanning system can be broken down to the following three tasks according to the state chart diagram shown in Figure 17.3.

1. select direction and start moving;
2. stop moving;
3. select mouse action or request further movement.

The task hierarchy using a CPM-GOMS model of a sample session is shown in Figure 17.4.

The simulator can determine the optimal direction for movement from the source and target coordinates. If a horizontal or vertical line from the source can reach the target, then one of the left, right, up or down directions is chosen. Otherwise, the pointer is moved diagonally until it reaches the same horizontal or vertical line as the target.

The eight-directional scanning system takes the scan delay (the time interval between any two state changes of the system) and scan step (the distance crossed by the cursor in a scan delay) as input. The default values of the two parameters are 1 sec and 10 pixels respectively.

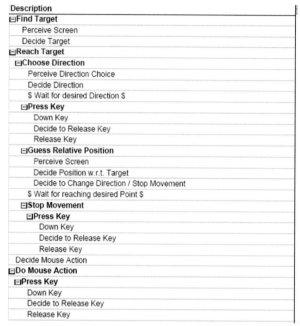

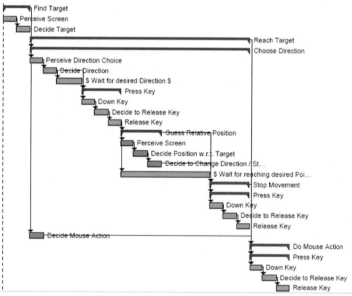

Figure 17.4. CPM-GOMS analyses for a clicking on a target using eight-directional scanning

Modelling Sub-optimal Behaviour

In the eight-directional scanning system, we have found users behave sub-optimally for the following three reasons:

1. they do not always choose the best direction of movement;
2. they do not stop the pointer movement at the correct place;
3. they try to place the cursor exactly over the centre of the target before clicking.

We have modelled the sub-optimal behaviour by a rule-based system developed in CLIPS (Version 6.24 of CLIPS, 2007). The rule based system models the uncertainty in choosing a direction of movement and the stopping position of the pointer. The rules for choosing direction take the difference in target coordinates and current coordinates as input and give the probabilities of different direction choices as output. The eight-directional scanning system shows the direction choices in a particular sequence. The general structure of a rule to select a direction is as follows:

Compare the difference in X and Y coordinates

> *Choose optimum direction choice with probability p_1*
>
> *Choose direction choice that comes after the optimum direction choice with probability p_2*
>
> *Choose direction choice that comes before the optimum direction choice with probability p_3*
>
>> *Where $p_1 > p_2 > p_3$ for UpLeft, UpRight, DownLeft and DownRight*
>>
>> *And $p_1 >>> p_2 > p_3$ for Left, Right, Down and Up, since novice users prefer Manhattan direction choices to diagonal*

The values of p_1, p_2, and p_3 are chosen based on observation of users' interactions and also a little bit of trial and error to get a good fit for the prediction. However the parameters were kept the same for all participants to generalize the model. We have found that users show more sub-optimal behaviour as the separation between source and targets increases. So the probability of correct direction choice is made inversely proportional to the separation between source and target. Since CLIPS fires rules concurrently, a direction choice may appear with two different probability values. In that case we consider the average probability.

We have found that users stop the pointer movement almost optimally when the source to target distance is small (less than 700 pixels for a 1,280×800 pixel resolution screen) or when the distance is very large (more than 1,300 pixels). In other situations, users often stop the pointer movement before or after the optimum instant. When the pointer is close to the target, then users also frequently fail to stop the pointer at the optimum point. So the rules take the source to target distance as input and give the probability of deviation of input from optimum position as output. The general structure of rules for stopping is as follows:

Consider the distance from the target

> *deviation_of_input is 0 with probability p_1*
>
> *deviation_of_input is -1 with probability p_2*
>
> *deviation_of_input is 1 with probability p_3*
>
> *deviation_of_input is -2 with probability p_4*
>
> *deviation_of_input is 2 with probability p_5*
>
>> *where $p_1 >> p_2, p_3 > p_4, p_5$*

In the fragment above, *deviation_of_input* 1 means the pointer is stopped after going one step further than the optimum stopping position, similarly *deviation_of_input* −1 means the pointer is stopped one step before the optimum stopping position.

17.2.3 Evaluation

The model of sub-optimal behaviour was validated with able-bodied users, who had not used this scanning system previously. In this case the participants behaved sub-optimally because they were novices. In our experiment, the participants were instructed to select buttons placed randomly on a screen. The buttons had to be pressed in a particular sequence, chosen randomly for each execution of the experiment. The random arrangement of buttons ensured that the experimental set up was not biased towards any particular screen layout or navigation patterns. All of the buttons were coloured red except the next target, which was green. After each button press, the last pressed button was disabled to show that it was no longer a target. The buttons were also labelled with a serial number to indicate the sequence. The actual task to be done by the scanning system was kept very simple so that it could not impose any cognitive load on the user. Hence any sub-optimal behaviour occurred only because of the scanning technique itself. The experiment was carried out in a Laptop with a LCD screen of resolution 1,280×800 pixels using the Windows XP operating system. A single keyboard switch was used to control the scanning techniques. The scan delay was set at 1,000 ms. The dimension of the buttons was 25×40 pixels and kept constant throughout the experiment. Eight able-bodied participants undertook the experiment, undergraduate and graduate students of the experimenters' institute. None of them had any colour-blindness: six participants were male and two were female. Their ages ranged from 23 to 35 years. The actual and predicted task completion times are shown in Table 17.1. The predictions are obtained by running Monte-Carlo simulation. It can be seen from Figure 17.5 that, with two exceptions, the model can predict task completion time with an overall standard error less than 3%. We also do not find any statistical significant difference between actual and predicted task completion time (t = 0.31 for a two-tailed paired t-test).

Table 17.1. Actual and predicted task completion time (in sec)

Participants	Actual	Predicted	Difference
P1	364	384	5.5%
P2	391	506	29.4%
P3	386	370	4.2%
P4	367	457	24.5%
P5	314	335	6.7%
P6	303	303	0.0%
P7	299	312	4.4%
P8	473	474	0.2%

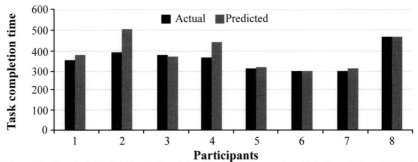

T(Actual, Predicted)=0.31, Std. Error (Actual, Predicted)=10.58 sec., Aver. Task Compl.Time=362.125 sec.

Figure 17.5. Comparing actual and predicted task completion time

17.2.4 An Application of the Simulator

In both Cartesian and polar scanning systems, the interaction rate of users remains very low. We have developed a new scanning technique based on clustering the screen objects, and have used our simulator to compare its performance with eight-directional and block scanning systems (which iteratively segment the screen area into equal sized sub-areas). We model only the scanning system, not the primary task done by it, so in this case we used cursor traces captured from interactions by able-bodied users as the input data. It has been found that our cluster scanning system can outperform other scanning systems. Further details about this study can be found in a separate paper (Biswas and Robinson, 2007).

17.3 Conclusions

In this paper, we have described a simulator that has been developed to predict the time and possible interaction patterns for disabled users undertaking a task. The simulator works for users with different levels of skill and physical disabilities. In particular we have confirmed the accuracy of the simulator for eight-directional scanning. We have also used the simulator to compare two other scanning systems. The experiment with able-bodied persons has confirmed the correctness of the model for novice users. Our next step is to populate the remaining components of the models with more details and to validate them with some experiments with people with disabilities.

We would like to thank the Gates Cambridge Trust for funding this work. We are also grateful to the students of Computer Laboratory and Trinity College, Cambridge for taking part in our experiments.

17.4 References

Biswas P, Bhattacharya S, Samanta D (2005) User model to design adaptable interfaces for motor-impaired users. In: Proceedings of the IEEE Tencon'05, Melbourne, Australia

Biswas P, Robinson P (2007) Performance comparison of different scanning systems using a simulator. In: Proceedings of the 9th European Conference for the Advancement of the Assistive Technologies in Europe (AAATE 2007), San Sebastián, Spain

Byrne MD (2001) ACT-R/PM and menu selection: applying a cognitive architecture to HCI. International Journal of Human Computer Studies, 55(1): 41–84

Fitts PM (1954) The information capacity of the human motor system in controlling the amplitude of movement. Journal of Experimental Psychology, 47: 381–391

John BE, Kieras BE (1996) The GOMS family of user interface analysis techniques: comparison and contrast. ACM Transactions on Computer Human Interaction, 3(4): 320–351

Keates S, Clarkson J, Robinson P (2000) Investigating the applicability of user models for motion impaired users. In: Proceedings of the 4th International ACM Conference on Assistive Technologies (ASSETS'00), Arlington, VA, US

Keates S, Trewin S (2005) Effect of age and Parkinson's disease on cursor positioning using a mouse. In: Proceedings of the 7th International ACM SIGACCESS Conference on Computers and Accessibility (ASSETS'05), Baltimore, MD, US

Kieras D, Meyer DE (1997) An overview of the EPIC architecture for cognition and performance with application to human-computer interaction. Human-Computer Interaction, 12(4): 391–438

Kieras DE (2005) Fidelity issues in cognitive architectures for HCI modelling: be careful what you wish for. In: Proceedings of 11th International Conference on Human Computer Interaction (HCII 2005), Las Vegas, NV, US

Lesher GW, Rinkus GJ, Moulton BJ, Higginbotham DJ (2000) Logging and analysis of augmentative communication. In: Proceedings of the RESNA Annual Conference, Orlando, FL, US

Luck SJ, Chelazzi L, Hillyard SA, Desimone R (1997) Neural mechanisms of spatial selective attention in areas V1, V2, and V4 of macaque visual cortex. Journal Of Neurophysiology, 77(1): 24–42

MacKenzie IS (2003) Motor behaviour models for human-computer interaction. In: JM Carroll (ed.) HCI models, theories, and frameworks: toward a multidisciplinary science. Morgan Kaufmann, San Francisco, CA, US

McMillan WW (1992) Computing for users with special needs and models of computer-human interaction. In: Proceedings of the ACM CHI 92 Conference on Human Factors in Computing Systems (CHI 92), Monterey, CA, US

O'Neill P, Roast C, Hawley M (2000) Evaluation of scanning user interfaces using real time data usage logs. In: Proceedings of 4th International ACM Conference on Assistive Technologies (ASSETS'00), Arlington, VA, US

Petrie H, Hamilton F, King N, Pavan P (2006) Remote usability evaluations with disabled people. In: Proceedings of the ACM CHI 2006 conference on Human Factors in Computing Systems (CHI 2006), Montréal, Canada

Reynolds JH, Desimone R (1999) The role of neural mechanisms of attention in solving the binding problem. Neuron, 24: 19–29 and 111–125

Rizzo A, Marchigiani E, Andreadis A (1997) The AVANTI project: prototyping and evaluation with a cognitive walkthrough based on the Norman's model of action. In: Proceedings of Symposium on Designing Interactive Systems: Processes, Practices, Methods, and Techniques (DIS97), Amsterdam, The Netherlands

Stephanidis C, Paramythis A, Sfyrakis M, Stergiou A, Maou N, Leventis A et al. (1998) Adaptable and adaptive user interfaces for disabled users in the AVANTI project. In: D Ranc, IP van der Bijl, S Sedillot (eds.) Intelligence in services and networks: technology for ubiquitous telecom services. LNCS-1430, Springer, Berlin/Heidelberg, Germany

Version 6.24 of CLIPS (2007) Available at: www.ghg.net/clips/Version624.html (Accessed on 21 May 2007)

Chapter 18

User-led Design of Technology to Improve Quality of Life for People with Dementia

R. Orpwood, J. Chadd, D. Howcroft, A. Sixsmith,
J. Torrington, G. Gibson and G. Chalfont

18.1. Introduction

18.1.1 Background

There has been quite a flowering of interest in exploring the design of assistive technology to support people with dementia. Just as with any other disability, there is a lot of scope for assistance through items of technology. Areas explored have varied from quite straightforward simple devices up to full-blown autonomous smart home installations (Orpwood, 2006; Bjoerneby, 1997). However, nearly all this work has aimed at improving the safety and security of people with dementia, and its impact on their quality of life has been a by-product (Marshall, 2001). This project reflected the need to address quality of life issues more directly through the design of appropriate technology (Orpwood *et al.*, 2007).

The project was a multi-disciplinary one involving the Universities of Liverpool (Social Science), Sheffield (Architecture) and Bath (Engineering). It also included representatives from Northamptonshire Social Services, Dementia Voice, and Huntleigh Healthcare, as well as Sheffcare and the Research Institute for the Care of Older People (RICE). It was funded under the EPSRC Equal programme and built on previous work undertaken by the three principal researchers involved. A lot of quality of life work involving assistive technology has tended to start with items of technology, and then to use various measures to assess its impact on the user's well-being. This project turned that process on its head, and started with people with dementia and a survey of what quality of life meant to them. Key issues were teased out from interviews, and this information then formed the basis of the technology design.

18.1.2 User Survey

The project started with a user survey involving people in the early to moderate stages of dementia, 16 living in their own homes, and 10 in care homes (Sixsmith *et al.*, 2007). The interviews were loosely structured, and the outcomes transcribed and analysed using a grounded theory approach to pull out the issues that were of importance to the users, as far as their quality of life was concerned. Initial workshops led to a wish-list of the key issues that were important to the quality of life of people with dementia (Table 18.1). In all the discussions that took place during the compilation of the wish-list there was little talk about specific items of technology. We were keen for the work to be led by user-needs rather than by any concern about technological feasibility.

Once the wish-list had been compiled, subsequent workshops explored possible technological solutions that could be applied to these issues. A total of 69 were considered, and there was strong feeling that there were many more possibilities that could have been included. This potential for the use of technology was very encouraging. The list was configured according to priority, as judged from criteria established by the consortium. Four items were selected from this list for some prototype development work within the project.

Table 18.1. Wish-list of issues to improve quality of life

Theme	Description
1. Oral/personal histories	Promoting reminiscence both when alone or with others
2. Social participation	Assisting people with forming new or continuing old relationships with family and friends. Encouraging and assisting with family visits
3. Conversational prompting	Supporting the act of conversation with others
4. Encouraging use of music	Promoting the enjoyment and use of music, either through playing or listening
5. Encouraging community relationships	Promoting relationships, and helping participation with local community
6. Supporting sequences	Supporting activities that involve a series of steps
7. Exercise/physical activity	Encouraging and supporting physical activity
8. Encouraging access to nature	Encouraging and assisting with access to outdoor spaces and nature
9. Sharing experiences of care and caring	Providing support with physical care tasks to free quality time between carer and PWD
10. Creative activities	Supporting people to take part in hobbies, pastimes and creative activities
11. Pottering in the home	Participation in minor tasks and household chores

18.1.3 Design Approach

The team in Bath has had quite a bit of experience of designing AT for people with dementia, and has evolved methodologies that seem to help with this process (Orpwood et al., 2003, 2005). One of the key features of the methodologies was the need for equipment to build on the sense of familiarity. In other words in the design of technologies such as supportive cookers, the cooker should look and feel just like those that were familiar to users, and the way that it dealt with any misuse should not involve anything that was outside common experience of cooker usage. Although this approach was useful for the INDEPENDENT project, there was also a need to explore control interfaces that were intuitive for people with dementia. These interfaces weren't just following what was familiar, but looked more fundamentally at what constituted an intuitive control. This approach meant the design work had to be very exploratory and iterative in nature so as to follow any clues that were provided from observation of the way the clients behaved.

18.2 Music Player

18.2.1 Initial Approaches

One of the issues which came out quite strongly from the user survey was the importance of music to the well-being of the interviewees. There appears to be a heightened emotional response on the part of people with dementia that raises the importance of music in their lives, and it is also a potent reminder of past events. However people with dementia find dealing with music-playing equipment confusing, and as with other aspects of their lives that remind them of their reducing abilities, they tend to withdraw from even trying to use them. In care settings the CD player is likely to be put on by care staff, and they will choose what to play. Residents don't feel any sense of involvement. We decided to explore the design of a simple music player that could be accessed by someone with dementia, and provide them with choice as to what they listened to.

The incorporation of a variety of controls, and the packaging into ever smaller enclosures, seems to be part of the general approach of designers of music playing equipment. It was reasoned that for most people, having adjustment of volume, tone, balance, etc wasn't necessary once it had been set up. The mixing of several music playing media such as radio, CD, tape etc into one package also just added to the confusion. Designs were therefore explored where the user control was reduced to just one or possibly two buttons. All the other controls which are infrequently used, such as volume, would be accessible to carers but hidden away so as not to confuse the demented user.

The iterative design work was enabled through the collaboration of the Research Institute for the care of the Elderly in Bath who were able to provide contact with a large number of local people with dementia who were happy to try out prototype devices. All these early exploratory tests were carried out with supervision in users own homes, over a period of half an hour or so.

18.2.2 Iterative Design Work

The initial music player was a CD player with one on/off button only (Figure 18.1a). The user had to insert the CD, and then just press the button. It assumed that users were familiar with CDs, and able to manipulate them into the drive, but it was found that users became very confused about which way up the CD should go, and about the process of opening up a lid and closing it again. However the main difficulty with the CD player was the time lag that occurred between pressing the play button and music coming out. This was only a second or two, but enough to cause a lot of concern. Users would press the button more than once, thereby turning it off again, or they would think they had done something wrong and take the CD out again. This seemed such a simple problem but it really made the player very difficult for many to use.

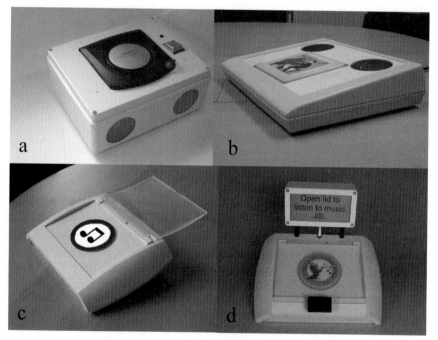

Figure 18.1. Stages in the evolution of the music player

It was decided to design a music player that used solid state recordings, like an MP3 player. The music would immediately play when the play button was pressed. The initial design just explored the play function, and ended up with a large round illuminated button that the user simply had to press and the music started (Figure 18.1b). This action was quite intuitive to the users. The original player had two small speakers mounted on the top of the player surface as well as the large button. Some users were confused about the round speakers, and tried to press these rather than the button. It was clear that there had to be no ambiguity arising from the appearance of the player surface, so that there was only one obvious choice for the user. Also the button clearly didn't stand out for some users despite its bright colour, and a simple modification was carried out to raise it a few millimetres above the surface of the player. This action worked very well. Tests were carried out over a period of weeks, so most of the users would have forgotten they had even seen the player when we returned with modifications. So their reaction to the controls was not really a learnt response. It had to be very obvious and unambiguous what the user had to do.

In parallel with this control work some further tests were carried out to try and find some image that conveyed to the user the idea of music. We wanted to be able to package the whole device in some form that conveyed the "I play music" identity. A series of images were put on cards, and users were asked what the images conveyed to them. Some of these were pieces of equipment, and some were images such as musical instruments. There was no one shape that clearly seemed to convey the musical function, other than things that looked generally like radios. But there were clear indications that musical notation for many people conveyed the idea of music, as did images of people playing musical instruments. It was decided that the player would be packaged in as simple an enclosure as possible so as not to be distracting, but that the control would be decorated using the musical identifiers we had found.

18.2.3 Choice of Music

The initial prototype device just explored the play function, but we were keen to provide an element of choice. A large number of possible ways were explored to achieve this end without increasing the complexity of the interface. The ideas considered tended to fall into one of two categories; designs that used a series of buttons, each of which activated a different set of music, and designs that used a single control that cycled through tunes. It was felt that the multitude of buttons would be confusing so cycling was explored to see how users got on. Initially the large button was configured so that every time it was pressed the music changed character. It was a bit like changing a CD. Users would press the button until they liked what they heard and then they would stop. This process seemed to be very intuitive. However it did leave the problem on an on/off control. We didn't want to have an extra button so we incorporated a clear plastic lid over the button. The button could be seen but the lid had to be lifted to access it (Figure 18.1c). The lid lifting provided the on/off control, and seemed to work well.

We made an interesting observation during these tests. The first player used the lid to turn the device on. The user then had to press the button to provide music, and press it again to change the type of music. To turn the whole thing off they had to shut the lid. We found with several users that when we asked hem to turn off the player they carried on pressing the button as though this was what was needed. However, it was notice that if they didn't think about what they had to do, if they were distracted for example, or their attention was drawn to something else and they had to stop the music to attend to it, without really thinking they would shut the lid. Lid opening and shutting seemed to be an intuitive turning on and off function, but when we were asking them to think what they had to do, they became anxious and confused. The new device used lifting of the lid to start the music and shutting to stop it. The button was purely used to cycle through the tracks. This worked very well.

18.2.4 Final Design

All the work up to this point had been carried out in user's homes and carefully supervised. Once we had achieved a design that seemed to function well a couple of units were made available for unsupervised tests with our colleagues in Liverpool and Sheffield, and which incorporated logging devices so that we knew when the device was used and at what time. Usage was fine initially, but several of the users eventually forgot about the existence of the player and stopped using it. A simple modification was made whereby a small illuminated panel was attached to the player. The panel would light up for a couple of minutes every half hour, and then turn off again (Figure 18.1d). This attention-drawing feature worked really well.

The final form of the device does not look particularly remarkable, but the way people with dementia interact with such equipment is not at all obvious, and requires the kind of careful iterative process that has been described.

18.3 Window on the World

The window on the world device was aimed at trying to reduce social isolation by providing a means whereby people could bring images of the outside world into their homes. In one home with a networked TV system with all the terrestrial channels, together with CCTV images from a security camera in the foyer, it was the CCTV channel that people watched! The camera was there to provide residents with a view of any visitors, and they particularly liked people watching. The initial window on the world device was simply a video camera with a radio link to a monitor in the home. Cameras were tried in different positions, and although some residents liked views of the garden to watch birds *etc.*, the main attraction was watching people passing by. A choice of cameras was made available through a handheld remote, but although residents found it easy to use it was often mislaid, or the residents felt it wasn't "theirs", and so they shouldn't use it. Touch screen activation was much more effective.

An extension of this basic set up used a two-way system with cameras and monitors in both the home of someone with dementia, and another in their carer's home. The idea was not to show talking heads in the manner of video phones, but rather to provide a kind of virtual visit to the carer's home (Figure 18.2). So the camera gave a wide view inside a room so that activities could be watched. Conversation was possible but the main aim was one of a kind on immersion in the carer-home environment. One of the issues that had to be dealt with was again that of intuitive controls. We wished to emulate the kind of social etiquette required when visiting someone. So we needed a virtual "knock-on–the" door facility, and a means whereby the carer could say "please come back later", and to say "it is time to go now". Our positive experience with touch screens for the one-way system encouraged us to use this for the two-way one. It has been our finding, and that of others, that touch screens are very intuitive to use by people with dementia. The system was tested in supervised situations and worked very well. Tests in users home are currently underway.

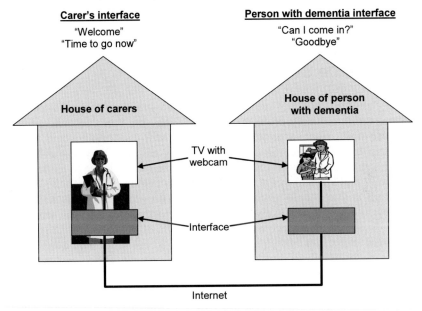

Figure 18.2. Two-way window on the world

18.4 Conversation Prompter

People with dementia find conversation difficult because of their memory problems and the likelihood that they will loose their train of thought whilst talking to someone. We aimed to provide a simple device that we called a help-me box which just repeated the last few works spoken. If someone lost their train of thought they pressed the button on the box, and the last five seconds or so of what they were saying would be repeated. It was hoped that this would enable them to recover what they were trying to say. Tests with Wizard of Oz techniques worked quite well where the experimenter sat in an adjoining room, and provided the reminder without having to develop any technology, just to see what sort of responses worked. A solid state recording device was developed but it was found necessary to incorporate a delay to take account of the decision time on the part of the user, otherwise you just replayed the non-verbal sounds that occurred whilst they were deciding to press the button. Tests on the prototype system were disappointing however. We recorded many conversations with people with dementia, and transcribed and analysed why they were breaking down. More often than not the speaker would try to cover up the fact they had lost their train of thought, and the help-me device was not really an appropriate solution. One of the main conclusions from this work was that the most important role for technology in this area is to provide prompts and reminders about topics of conversation. Some means of structuring these prompts so as to find areas of interest would be really useful. The conclusions of the CIRCA project showed just how effective reminiscence can be to encourage conversation, and improve carer/user relationships.

18.5 Sequence Support

Many tasks comprise a series of sub-tasks, such as making a cup of tea. For people with dementia their memory problems make dealing with these sequences very tricky because at any given moment you are engaged with a subtask but have to see its position in the whole activity. With tea-making the user may know that it involves boiling water but forgets when that should be done relative to putting tea in the teapot *etc.* All kinds of things can result such as putting tea bags in the kettle. The end result is that people tend to withdraw from even trying such activities because they know they are likely to fail, and it just reinforces their sense of uselessness. The user survey uncovered many aspects of people's lives that were clearly of great importance to their sense of well-being, but which would not stand out as being particularly important to cognitively able people. Being able to make yourself a cup of tea is typical of these kind of activities.

This project aimed to find ways of supporting sequences. The two main areas of work were; providing prompts for each step in a sequence, and knowing when one step is complete and moving on to the next. The work completed within the project just aimed at the first of these activities. All the testing was done using Wizard-of-Oz techniques where the researcher provided the decision that a step had been completed and moved things on.

A series of tests was completed looking at the best way of representing the stages of a task, and encouraging activities. Three test tasks were chosen; putting a cassette into a player, finding a programme in a TV listings, and putting a letter into an envelope and sealing it. All three were supported using a screen-based system (Figure 18.3) that explored four different means of providing the prompt.

These were the use of text, the use of photos, the use of video, and the use of voice. At the start we assumed the video would be the most supportive because it most closely reflected reality. This was not the conclusion however. It was interesting to observe the use of video and the photos. Participants were unable to generalise what the prompt was telling them. They would try to follow precisely the actions displayed. The letter had to be folded in exactly the same position as on the photo or the video. The envelope had to be in the same orientation as that shown in the visual images. The cassette player used was the same one appearing in the visual images, but the view was not the same as that seen by the user. It was slightly higher and at an angle. These factors made the two visual media far less effective, and caused some consternation because they couldn't be followed exactly. On the contrary the text prompt was clearly quite effective. The user was following instructions rather than copying. The tests were repeated with the audio prompt in addition to the other three media, and this was found to improve the success further. It was also clear observing the users that a one-off reminder wasn't enough. They would want to go back to the prompt repeatedly before they could complete a subtask. The text made this easier because the whole action was there in front of them, and they could repeatedly check it. The combination of text and a one-off audio prompt seemed to work best in our tests.

Figure 18.3. Sequence prompting device

Although the moving on from one stage to the next in the sequence was controlled by the researcher, there were indications about more automated approaches that could be explored. It would be difficult to use video processing to discern when a sub-task was complete, although some work has been dome in this area. It was felt that a simpler approach would be either for the user to make the decision and press a "next" button, or the system would need to prompt with a "have you done that" message to get a yes/no response. From the work we completed on the music player it is clear that a lot of iterative user work would be necessary to find an optimum design for such a system. However, the key component of the device, the means of breaking up tasks into subtasks and providing a prompt, was very successful.

18.6 Conclusions

The project has concluded with both a wish list of topics that need addressing in order to improve the quality of life of people with dementia, and a list of potential technologies that could achieve some of these. The four items explored were variously successful but some useful design guidelines were concluded, particularly in relation to the design of control interfaces that are intuitive to people with dementia. These are listed below.

- Keep things very simple, preferably just one control.
- Ensure the control stands out from the background and there are no distractors; even a screw head can be interpreted as something to be pressed!
- Cause and effect need to be clear; even short time delays can cause confusion.
- Pressing seems a lot easier than other forms of interaction, such as toggle switches.
- Controls need to be large and provide some feedback that something has been initiated, such as a feel-able or audible click, or the turning on of illumination. Non-moving touch switches are very ineffective.
- In order to test whether a control is intuitive or not, the user needs to be in a position of acting automatically without too much thinking. If you simply ask them to do something, their anxiety may overshadow their response.
- Opening and shutting seems to convey start and stop quite effectively.
- Touch screens appear to be a really intuitive kind of control interface.
- Prompts seem to be more effective in the form of basic verbal or text instructions, rather than trying to encourage copying.
- Musical notes and people playing musical instruments both seem to covey the idea of music well.

It is hoped that others keen to use assistive technology to improve quality of life of this important group of people can build on this work and the conclusions we have made.

18.7 Acknowledgements

We would like to thank the EPSRC for the funding for this work through their EQUAL programme, and for the many people with dementia and their carers who helped guide the progress of the project.

18.8 References

Bjoerneby (1997) The BESTA flats in Tonsberg. Using technology for people with dementia. Human Factors Solutions, Oslo, Norway

Marshall M (2001) Dementia and technology. In: Peace S, Holland C (eds.) Inclusive housing in an ageing society. Policy Press, Bristol, UK

Orpwood R, Faulkner R, Gibbs C, Adlam T (2003) A design methodology for assistive technology for people with dementia. In: Craddock G, McCormack L, Reilly R, Knops H (eds.) Assistive technology – shaping the future. IOS Press, Amsterdam, The Netherlands

Orpwood R, Gibbs C, Adlam T, Faulkner R (2005) The design of smart homes for people with dementia – user interface aspects. Universal Access in the Information Society, 4: 156–164

Orpwood R (2006) Smart homes. In: Pathy M, Sinclair A, Morley J (eds.) Principles and practice of geriatric medicine, 4th edn. John Wiley and Sons, Chichester, UK

Orpwood R, Sixsmith A, Torrington J, Chadd J, Gibson G, Chalfont G (2007) Designing technology to support quality of life of people with dementia. Technology and Disability, 19: 103–112

Sixsmith A, Gibson G, Orpwood R, Torrington J (2007) Developing a technology "wish list" to enhance the quality of life of people with dementia. Gerontechnology, 6: 2–19

Chapter 19

Photonote: The Making of a Classroom Adaptation System

G. Hughes and P. Robinson

19.1 Introduction

Taking accurate notes during a lecture can be a difficult task for many students with disabilities due to the visual nature of the material, the oral nature of the lecture, and/or the physical act of taking notes. By utilising high-resolution digital cameras it is possible to capture the contents of a whiteboard or overhead projector to allow future presentation to a student with a disability, without requiring the lecturer to adapt to new technologies (Hughes, 2006). Our system, called Photonote, combines enhanced high-resolution imagery of displayed information with video of a lecturer and sign-language interpreter to provide a lecture revision system for students with disabilities. Currently these students employ many methods to compensate for their disability, including university-appointed note-takers, copies of friends' or lecturers' notes, or simply aiming to teach themselves the material from textbooks. This paper discusses the creation of the Photonote system, the evaluation of the system, and notably, the qualitative analysis of the study.

19.2 Related Work

Various systems have been designed to capture and record lectures. However, none have been designed specifically to aid students with disabilities. The eClass project was started at the Georgia Institute of Technology with the initial goal of creating a "classroom environment in which electronic notes taken by students and teachers could be preserved and augmented with audio and video recordings" (Abowd, 1999). Using a 72" touch screen display, called a LiveBoard, a professor could write notes into proprietary software called ZenPad. This system required training, and even the researchers of eClass (Abowd, 1999) admit that lecturers generally use whiteboards three times the size of the 72" LiveBoard which their system implements. The Lecture Browser (Mukhopadhyay and Smith, 1999) system was designed to process a lecture for later viewing. However, much like eClass's LiveBoard system, this requires the lecturer to be aware of the system and to upload PowerPoint files to a server after the lecture has finished. AutoAuditorium (2005) is a commercial system designed to automate the lecture-recording process. In its simplest form it consists of two static cameras and one automatic pan/tilt camera. The pan/tilt camera is used to follow the presenter during the lecture, one camera is used to capture an overall view of the theatre, while a third camera is used to capture the output of a digital projector. This system does not require the lecturer to be aware of its presence. However, it is designed to create an aesthetically-pleasing video rather than provide assistance for students with disabilities. Apreso (Anystream Incorporated, 2007) is a popular lecture-recording system used by many universities including Temple University, the University of Pittsburgh, Princeton University, and the University of Tennessee. This system can capture presented material using many different methods. The primary method forces the lecturer to use a computer to present visual information. Apreso can also be configured to capture visual information with a standard-definition video camera. However, standard definition video is not capable of capturing enough detail to read presented visual information (See Section 19.4.1).

Numerous other lecture-recording systems are being researched and are available commercially; such as Lectopia (Arts Multimedia Centre, 2007) and Tegrity (Tegrity Incorporated, 2007). There are two primary differences between Photonote and any other system, either in development or in production. First, Photonote is designed to be able to capture intricate visual information without requiring lecturers to adapt their normal manner of presentation to the system. Second, Photonote is designed specifically to address the needs of students with disabilities, a large group of students neglected by the majority of lecture recording systems.

19.3 Existing Solutions

Universities acknowledge the problems that students with disabilities encounter in a lecture environment and employ many methods in an attempt to aid students with disabilities. Methods vary widely, based upon the resources available to the university and also on the individual needs of a student. There is no single solution that works for everyone. However, with a variety of choices available, universities try to find a solution that is acceptable for each individual.

Universities commonly appoint a note-taker to aid an individual with a disability which impairs their ability to take notes. This note-taker can either be another lecture attendee who is paid for the notes or a professional note-taker who is only attending the lecture to take notes. The former is more common because it is far cheaper to pay a student rather than a trained note-taker. Following the lecture, the note-taker's notes are photocopied and distributed to any individuals with disabilities who have difficulty taking their own notes. Students with disabilities normally receive a copy of the lecture notes within a few days of the lecture. Despite its advantages, this method presents many problems. The largest problem arises from the time-shifting of information. Visual information presented during a lecture is usually intended to be accompanied by the spoken lecture. Therefore, the lecture is useless without the accompanying visual information, and the visual information alone is useless without the accompanying oral presentation. Problems can also arise from errors in a note-taker's notes or artefacts from photocopying.

A university may also hire a dedicated note-taker for an individual with a disability which impairs their ability to take accurate notes. Instead of providing photocopied notes for a number of students with disabilities, this note-taker would take notes and hand them directly to the student during the lecture. This solution prevents the time-shifting problem of most note-taker solutions as well as providing the individual with a disability the ability to take his/her own notes regarding the presented information. This solution can be distracting to other lecture attendees as verbal clarification of words or symbols may be necessary and the exchange of paper between the two individuals may also be distracting. This is also a costly solution since there must be a one-to-one ratio between note-takers and students with disabilities.

Another method of obtaining accurate lecture notes is for a lecturer to provide a written copy of the information which will be presented during the lecture. These notes may be copies of overhead slides, printouts of a PowerPoint presentation, or even photocopies of hand-written notes. The time-shifting problem encountered when using a note-taker is avoided because a lecturer's notes can be given to a student prior to the lecture. Lecturer's notes are less likely to have errors in them, but they are not necessarily problem free. These notes can be very difficult to read if they are handwritten. Additionally, a lecturer may only write personal cues regarding the information to be presented rather than complete notes. Another drawback to this method is that lecturers often choose not to follow their notes directly, skipping a portion here or there. An individual with a visual disability might have difficulty following the notes if they assume that the order of the notes and spoken presentation will coincide. Furthermore, receiving a copy of a lecturer's notes is not particularly beneficial to someone with a hearing impairment who is watching a sign-language interpreter. Such students have access to the presented visual information, but during a lecture their visual channel is occupied by the lecture translation provided by a sign-language interpreter or lip-reading.

19.4 Designing Photonote

The primary goal of Photonote was to develop a system that was able to replace human note-takers whilst also overcoming the difficulties of existing classroom-adaptation methods. Photonote is designed to capture and redisplay all of the information presented in a lecture for students with disabilities, without requiring lecturers to change their presentation techniques. The system operates unobtrusively, distinguishing it from many other lecture recording systems which require the presenter to use specialised hardware or undergo system training.

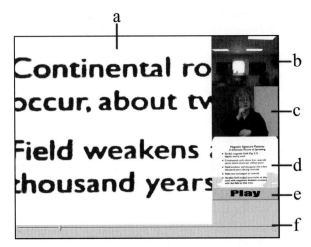

Figure 19.1. A screenshot of the Photonote application. (a) visual information window, (b) video of the lecturer, (c) video of the sign-language interpreter, (d) thumbnail of the visual information, (e) play/pause button and (f) time slider

The Photonote software (see Figure 19.1) displays three key pieces of information simultaneously. The largest area within the application displays a high-resolution, enhanced image of the visual information presented during the lecture. Also displayed are two videos, one of the lecturer and one of a sign-language interpreter, if present. Audio from the lecture is replayed for the student through a pair of standard headphones. All four of these sources are synchronised in time. A user can play/pause the videos as well as jump to any point within the lecture. The visual information window can be scrolled and zoomed by the user whilst the software is replaying the lecture. The system can be used live in a classroom or as a revision tool. This paper focuses on the use of Photonote as a revision tool rather than on the benefits that the system would provide to a student during a lecture.

19.4.1 Capturing Information

Existing lecture recording systems use many methods of capturing information, including digital whiteboards, customised software packages and standard video cameras. Since one of the main goals of Photonote was to design a system which would allow a lecturer to present normally, it was clear that using cameras would be the only method of capturing information in an unobtrusive manner. Although video would be the obvious choice to capture the visual information, time-lapse digital photography was chosen after laboratory experimentation which determined that the resolution of a digital-video camera was insufficient to capture the details of a lecturer's presentation. Using a standard Snellen Eye Chart, it was determined that a standard-digital-video camera had a visual acuity of 20/100. Therefore a digital-video camera would need to use a five-times zoom, or be five times closer to an object than human with average vision in order to distinguish the same amount of detail. This would reduce the camera's field-of-view correspondingly, losing its ability to capture all of the presented information. Using the same method, we found that an 8 Megapixel (MP) digital-still camera has better (>20/20) visual acuity than an average human while still maintaining a very wide field-of-view. Therefore such a camera could be placed at any location in a classroom where a person with average sight would be able to see the presented information.

The lecturer and sign-language interpreter are captured by two separate digital-video cameras, while a digital-still camera is used to capture the visual information which is presented during a lecture. The digital-still camera used, a Canon PowerShot S80, is able to capture one frame every three seconds which provides ample frequency, as presented information on a whiteboard or OHP will not change significantly within a three second time-span.

a   b   c

Figure 19.2. Demonstrates the enhancement process for a whiteboard within a classroom

19.4.2 Information Enhancement

Visual information can be enhanced for students with disabilities using a number of different techniques. In our system we employ computer-vision techniques to compensate for the angle at which the visual information is captured; then we enhance the contrast by running an adaptive threshold algorithm, and also remove obstructions and highlight new additions (See Figure 19.2). The final result of this process (Figure 19.1a) provides a very legible copy of the visual information which can be zoomed and changed to suit the needs of any individual using the system.

The visual enhancement begins with a perspective transformation which compensates for the angular difference between the camera and the source of visual information. After compensating for the angular difference between the information and the camera, whiteboards and OHP output can be further enhanced by using a modified version of the adaptive threshold algorithm developed by Wellner (1994) for use on the DigitalDesk. This algorithm traverses the image left to right and top to bottom while keeping track of the average colour of the last x number of pixels. When a pixel is encountered that is significantly darker than this running average, it is coloured black, while all other pixels are coloured white. The algorithm was modified slightly to exclude the dark pixel values from the running average. This algorithm helps to threshold an image that is not evenly lit: however, it requires that the image has more white space than dark space, such as a piece of paper, an OHP projection, or a whiteboard (see Figure 19.2b).

Temporary obstructions can be removed and new information highlighted by analysing the image after the adaptive threshold algorithm is executed. The image is traversed pixel-by-pixel, feeding each pixel into a blob-detection algorithm, along with a threshold. The blob-detection algorithm then searches surrounding pixels to find those with colour values within the given threshold. Each pixel within the threshold is marked as part of the blob, and then the pixel is fed as input back into the blob-detection algorithm. This method is recursive, ending when no more touching pixels are found with colour values within the threshold. The algorithm finally returns the total number of pixels in the blob, along with the blob's height and width. A threshold value is determined depending on the image being analysed. Blobs containing fewer pixels than this threshold value are considered to be text, and anything larger is considered to be a person, or some non-significant obstruction.

Obstructions can now be removed by analysing the large blobs and resorting to information from the same area in previous frames where no obstruction was present, which may require the system to check a number of different frames to find one without an obstruction. Although this enhancement can be useful to many students, it can present false information. If the lecturer was obstructing the whiteboard while erasing and writing new information the system would continue to display the old information. Highlighting and detecting the addition of new text is important because the user can lose spatial orientation while zoomed in on visual information and may be unaware of new information presented in other locations. By detecting new text the system can either automatically centre on the new information, or indicate to the user that new information has been added elsewhere.

19.5 Evaluation

To evaluate the effectiveness and usefulness of the Photonote system for students with disabilities, we conducted a user study. The goal was to determine if our electronic revision system could replace a human note-taker utilised by students with disabilities that impair their ability to take notes during a lecture. Our hypothesis was that students with disabilities would perform equally well, or better, on an examination when they used their own notes and Photonote to revise lecture material when compared to an examination where the students used their own notes along with the notes of a note-taker to revise. Additionally, feedback provided by participants will be crucial in developing future versions of the system.

The study had a repeated measure design where participants used two different methods to prepare for an examination. The 33 study participants were grouped according to their disability, or lack thereof. The disabilities being investigated included hearing impairments, visual impairments, learning disabilities, and mobility impairments. All participants attended two consecutive lectures on 6 January 2007. During the lectures, participants took notes as they normally would.

On 13 January 2007, a one-and-a-half hour revision period was held where students revised the two lectures using one of two methods. Method One entailed using the Photonote system to revise the material presented in a lecture along with the participant's own lecture notes. For Method Two, participants used their own lecture notes to revise, along with the notes of a note-taker if the participant had a disability. Participants had 45 minutes to revise each of the two lectures. A control group, consisting mainly of people without disabilities, used only their own notes to revise for the exam.

A written examination followed the revision period to test the students' retention and understanding of the information presented in the lectures. All odd numbered questions pertained to Lecture One, while even numbered questions pertained to Lecture Two. The odd questions will be considered Exam One, while the even questions will be considered Exam Two. We can define M_1 and M_2 to be the score of the exam that a participant took after utilising Method One and Method Two respectively. ΔM, defined as the difference between M_1 and M_2, will be positive for an individual if our system improved their exam score, and negative if our system decreased their exam score.

19.5.1 Quantitative Results

Participants did not perform significantly better using either of the two methods. A t-test and an ANOVA test both failed to provide any evidence that the use of the Photonote system, instead of a note-taker, had any statistically significant impact on a participant's test score ($p > 0.05$). These results show that our system would be a suitable alternative for some students with disabilities because it can effectively replace a note-taker, although it is not a universal solution. Further analysis was used to seek an explanation as to why the system helped some students but not others.

Information collected from participants' questionnaires was used to determine why some participants were aided by Photonote while others were not. Further statistical tests led to the detection of some interesting correlations. Most tests were inconclusive: however, it was found that there was an interesting correlation between ΔM and whether the participant normally uses a university-appointed note-taker. This is visible in Figure 19.3, which demonstrates that a person who habitually uses a university-appointed note-taker is more likely to benefit from the use of the Photonote system. This is a crucial finding as the original purpose of our system was to replace a university-appointed note-taker for students with disabilities. Combinations of variables were also examined, for example males with disabilities or females who have difficulty in school. One combination gave particularly interesting results: the combination of gender and whether a participant considers him/herself disabled. It was determined that our system was most likely to aid females without a disability and males with a disability. More details on the quantitative results of this user study can be found in (Hughes, 2007).

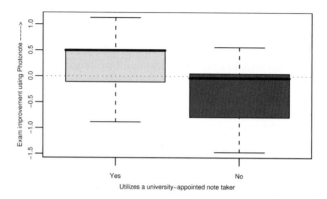

Figure 19.3. Box plot of change in exam results using Photonote versus utilisation of a university-appointed note-taker. Boxes encompass the range one standard deviation either side of the mean. The solid bar shows the median.

19.5.2 Qualitative Results

Following the user study a questionnaire was distributed to all participants which allowed them to provide feedback regarding the Photonote system. In general, the feedback was very positive. The participants were able to express which aspects of the system worked well for them and which aspects they had found the most useful. They also provided information regarding potential developments to the system which they thought would make it even more useful. For example, many individuals expressed interest in having the ability to enlarge the video of the lecturer and the sign-language interpreter. This feature was intentionally omitted from the tested version of Photonote to avoid confusion amongst participants because the video of the lecturer and sign-language interpreter cannot be zoomed or manipulated in the same way that the captured visual information can be manipulated. It would be possible to allow the user to resize the available areas to suit their own needs. For example, individuals with hearing impairments would have preferred a larger video of the sign-language interpreter while individuals with learning disabilities found the video of the sign-language interpreter to be distracting and would have preferred to eliminate that video entirely.

An individual with a hearing impairment suggested a close-up video feed of the lecturer for people who can lip-read. Standard-resolution video cameras are not capable of capturing a large field of view with enough detail to allow lip-reading of a subject at the front of a lecture theatre. Maintaining a large field of view is necessary because the entire front of a classroom must be captured to allow a lecturer to move about while giving their lecture. Capturing video with enough detail to allow lip-reading and still allow the lecturer to move about the classroom could be accomplished by using two cameras. One standard-resolution video camera could be used to capture an overview of the front of the lecture theatre and a computer-controlled pan-tilt-zoom (PTZ) camera would be used to capture details. Video from the single standard-resolution camera could be analysed to detect the location of the lecturer by using face-detection methods. The computer-controlled PTZ camera could then zoom in on the location of the lecturer's face, capturing ample detail to allow for lip-reading. Such solutions have been investigated and implemented by numerous other researchers for various different applications (Xu and Sugimoto, 1998; Prince *et al.*, 2005; Bazakos and Morellas, 2007).

One commonly requested feature involved some method of cataloguing or indexing a lecture. Participants found that it was time consuming to search through a lecture using a time-slider. Providing automatic indexing could be easily accomplished by computing the difference between frames of visual information. Every time the visual information changes by x percent, a new key frame of the visual information can be marked. Additionally, the user could control the value of this threshold, allowing them to control how many key frames are shown.

Many participants suggested the addition of an area where they could take their own notes. This was excluded from the study to force all users to concentrate on the lecture during the revision session rather than taking further notes. After all, this system is designed to help a student revise a lecture, not to provide a substitute for attending a lecture and taking one's own notes normally. Alternatively, the inclusion of a note-taking area may be a nice feature to add to Photonote. Aside from the obvious ability to take notes, such a feature would have many additional benefits. Notes taken by a user could be synchronised to the lecture and displayed at the appropriate time during future revisions. Notes would also provide further information by which lectures could be indexed and searched. Additionally, notes by students could be shared with other students via the network to allow collaboration during a revision session.

Participants provided many great ideas and suggestions along with very useful criticisms regarding the Photonote system. Their suggestions are currently being utilised to develop the next version of Photonote. It is expected that the next version will be able to address all of the problems that individuals identified including resizeable windows, a note-taking area, the ability to disable the sign-language interpreter video, and numerous other options. Future user studies will be needed to test whether the addition of the suggested features changes the usefulness of Photonote.

19.6 Conclusion

While looking at the numbers and statistical tests it is easy to forget that the participants are human and that it is very difficult to find universal solutions, especially when developing technologies for individuals with disabilities. Albeit the quantitative results are important, they don't tell the entire story as user feedback is a critical aspect of assistive technology development. The qualitative results of this user study will help to create a more useful and efficient version of Photonote.

The analysis suggests that our system could replace a note-taker for students with disabilities, but it is not a universal solution, just as the utilisation of a note-taker is not a universal solution for all people with the same disability. Some students were dramatically impaired when using our system instead of using their normal method of revision. However, our system proved to be very beneficial to other students, improving their performance in an examination. This first study conducted to test this software and lecture-capturing technique shows great promise for future revisions based on user feedback.

19.7 References

Abowd SD (1999) Classroom 2000: an experiment with the instrumentation of a living educational environment. IBM Systems Journal, 38(4): 508–530

Anystream Incorporated (2007) Apreso. Available at: www.apreso.com/ac_product_ overview.asp (Accessed on 12 July 2007)

Arts Multimedia Centre (2007) The University of Western Australia: Lectopia. Available at: http://ilectures.uwa.edu.au/ (Accessed on 12 July 2007)

AutoAuditorium (2005) Autoauditorium system home page. Available at: www.autoauditorium.com/ (Accessed on 16 August 2005)

Bazakos ME, Morellas V (2007) Face detection and tracking in a wide field of view. United States Patent Application Publication 20070092245

Hughes G, Robinson P (2006) Time-lapse photography as an assistive tool. In: Proceedings of the 3rd Cambridge Workshop on Universal Access and Assistive Technology (CWUAAT'06), Fitzwilliam College, University of Cambridge, Cambridge, UK

Hughes G, Robinson P (2007) Photonote evaluation: aiding students with disabilities in a lecture environment. In: Proceedings of the 9th International ACM SIGACCESS Conference on Computers and Accessibility (ASSETS'07), Tempe, AZ, US

Mukhopadhyay S, Smith B (1999) Passive capture and structuring of lectures. In: Proceedings of the 7th ACM International Conference on Multimedia (MULTIMEDIA'99), New York, NY, US

Prince SJD, Elder JH, Hou Y, Sizinstev M (2005) Pre-attentive face detection for foveated wide-field surveillance. In: Proceedings of the 7th IEEE Workshops on Application of Computer Vision (WACV/MOTIONS), Brekenridge, CO, US

Tegrity Incorporated (2007) Tegrity campus 2.0. Available at: //www.tegrity.com/ (Accessed on 12 July 2007)

Wellner PD (1994) Interacting with paper on the DigitalDesk. Technical Report, UCAM-CL-TR-330, Computer Laboratory, University of Cambridge, Cambridge, UK

Xu GG, Sugimoto T (1998) Rits eye: a software-based system for real-time face detection and tracking using pan-tilt-zoom controllable camera. In: Proceedings of the 14th International Conference on Pattern Recognition (ICPR'98), Brisbane, Australia

Chapter 20

FES Indoor Rowing and On-water Sculling

B.J. Andrews, D. Hettinga, R. Gibbons, S. Goodey
and A. Poulton

20.1 Introduction

Although individuals with spinal cord injury (SCI) regard exercise as important and clearly can benefit from proper exercise, there are several hurdles to overcome. Most importantly, it has been suggested that an exercise intensity of at least 6 METs (*i.e.* oxygen consumption of 21 ml/kg/min) is required to lower the relative risk for coronary heart disease (Tanasescu *et al.*, 2002) and to significantly improve blood lipids volumes of at least 1,200-2,200 kcal/week (Durstine *et al.*, 2001). However, many persons with SCI can have difficulty achieving these levels (Manns and Chad, 1999). Even though some can achieve moderately high peak oxygen consumptions using their upper body muscles alone, maintaining sufficient aerobic power with small muscle mass exercise is difficult. Exercise performance may be limited by local fatigue of the highly stressed arm musculature despite adequate systemic responses.

The solution may be functional electrical stimulation (FES), (www.FESrowing.org), of the paralyzed lower limbs to increase the amount of metabolically active muscle mass. In FES, muscles are stimulated to contract by applying a train of pulses to known motor points on the skin surface via adhesive electrodes. Although the voltage is fairly high (around 150V) the pulses are very short and have low energy, so are not harmful. However, FES exercise alone is not of sufficient intensity for many of the beneficial adaptations associated with aerobic exercise. Therefore, hybrid FES exercise that involves both innervated upper body and electrically stimulated lower body has been explored, and has been shown to produce significantly greater aerobic power and peak oxygen consumption than FES exercise alone (Verellen *et al.*, 2007). Wheelchair propulsion and arm-cranking ergometry have been associated with shoulder pain (Jacobs, 2004). Pulling actions have been proposed as therapy for chronic shoulder pain in wheelchair users (Jacobs, 2004). In chronic wheelchair users, this may help prevent upper limb overuse injury (Olenik *et al.*, 1995). FES rowing involves pulling actions and participants report it to be better tolerated than standard upper body exercise, even though the latter was conducted at a lower absolute oxygen consumption, VO_2 (Wheeler *et al.*, 2002).

20.2 FES-rowing Technology

20.2.1 Indoor Rowing

Andrews and colleagues have developed FES rowing with exercise intensities in excess of 35ml/kg/min with volumes greater than 2,000 kcal/week being achieved by some rowers with complete spinal cord injuries who have competed in international rowing competitions over the Olympic 2,000m distance alongside able-bodied rowers (Hettinga and Andrews, 2007). In November 2004 two paraplegics (Figure 20.1, subjects 1 and 2 in Table 20.1), using the 4-channel FES rower shown in Figure 20.2, successfully competed along with over 2,500 able-bodied rowers in Birmingham at the British Indoor Rowing Championships (BIRC), (www.concept2.co.uk/birc) – this event and the FES equipment is the subject of a permanent exhibit at the Rowing Museum (www.rrm.co.uk). Since then, paraplegics and quadriplegics have competed in the annual BIRC and the World Indoor Rowing Championships (WIRC or CRASH-B's). Their data are summarized in Table 20.1 which shows the level of injury on the standard ASIA impairment scale and their times for the 2,000m distance.

Table 20.1 Competitor data

Subject	1	2	3	4	5	6
Age (years)	50	34	44	28	18	22
Weight (kg)	70	65	77	52	51	90
Lesion, ASIA	T4 (A)	T8 (A)	T6 (A)	C7/8 (A)	C4 (A)	T12 (A)
Since injury yrs	5	5	6	13	2	2
FES-rowing training duration	18 months	12 months	3 months	14 months	7 months	12 months
FES muscle conditioning before rowing	3 months	>1 year ; included FES cycling	1 month	1 month	1 month	3 months
BIRC						
2004	12:02	13:59	na	na	na	na
2005	11:11	13:58	11:39	na	na	na
2006	11:12	na	10:35	na	25:13	10:28
WIRC 06	11:37	14:01	12:00	16:55	na	na

Figure 20.1. Paraplegic rowers at BIRC '04: subject 1 (co-author RG) in foreground

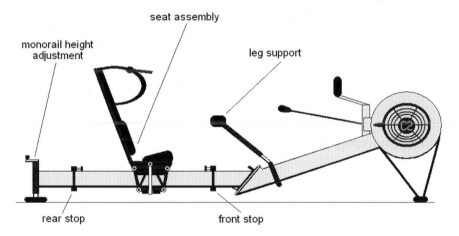

Figure 20.2. Schematic of adapted rower

The technology has been developed by a process of stepwise refinement based on feedback and suggestions from the rowers' experience. Mechanical adaptations include a modified seat assembly to support the trunk and a telescopic leg support to keep the knees together while allowing motion forwards and backwards. A footplate stabilises the feet, while spring loaded end stops limit the movement of the seat on the track. FES pulses are applied to electrodes on the quadriceps and hamstrings. The pulses are biphasic with a frequency of 10-100 Hz and pulse widths up to 500 μs. The drive and recovery phases of rowing are controlled using a changeover switch mounted on the handle. In the case of the quadriplegic subjects there is a changeover switch operated either by the attendant, or self-controlled by the rower, by positioning the switch so as to allow operation by wrist extension movements with or without velcro closed mitts around the handle grip.

Conventional Odstock stimulators have been used for much of the training, as they provide four channels which can be switched in pairs with manual adjustment of pulsewidth for each channel. Further improvements in performance and range of application may be made using computer controlled stimulators, which can support more complex control strategies. Ideally such a stimulator should allow simple programming by researchers who are not experts in real-time software for multi-tasking microprocessor systems, so that new strategies can be readily investigated. Ease of programming also facilitates the transfer of knowledge accumulated with the system to new researchers. Recent designs have been based on the Tiger range of processors (Wilke Technology GmbH) which run compiled multi-tasking BASIC. The stimulus pulses are derived by amplifying accurately timed digital outputs from the processor with high-current drivers feeding step-up transformers to give the required bi-phasic isolated signals. In the interests of safety, the equipment is battery powered and the channels are individually isolated to 1kV. While the Tiger processor provides the usual analogue and digital inputs which can be used for sensor feedback, we have also developed a distributed data acquisition system which uses an I2C bus interface. Position, acceleration and force data is acquired from sensors organised in clusters, each cluster having its own local processing provided by a PIC16F876 microcontroller (Microchip Technology, Inc.). Sensor types include accelerometers, rate gyroscopes, force-sensing resistors, bend sensors, and Hall effect position sensing devices. New strategies for open loop and closed loop control can thus be investigated readily (Poulton and Andrews, 2004).

20.2.2 On-water Sculling

Recently, we have further developed FES rowing in the tank and on-water. The Alden 16 (single) and Alden 18 (double) recreational rowing shells were chosen for stability with the removable stateroom module, Figure 20.3. A bungee cord was attached to the custom sliding seat to assist recovery as shown in Figure 20.3. To facilitate laboratory training and development a sculling simulator was developed based on the Alden stateroom module. Regular carbon fibre oars were cut down to remove the spoons and coupled to two 2-state hydraulic cylinders (2-state to closely simulate rowing *i.e.* minimal resistance during RECOVERY and adjustable resistance during DRIVE) as shown in Figure 20.4. In the simplest arrangement 4-channel surface stimulation was used (as with the indoor rower) with the control switch mounted on the oar handle. Training in the simulator followed a similar protocol to that used for rowing on the adapted Concept II shown in Figure 20.2.

Figure 20.3. The Alden (16 & 18), (www.rowalden.com), detachable state room with modified seat, padded shoulder straps and bungee cord, pulley and cam cleat to allow adjustment of bungee cord tension. The telescopic leg stabiliser is similar to that used on the indoor rower.

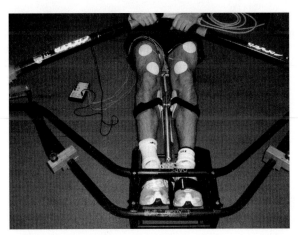

Figure 20.4. The FES rowing simulator, comprising an Alden boat stateroom fitted with hydraulic cylinders for laboratory training and development

Following at least two one hour sessions on the rowing simulator and prior to rowing on-water, FES rowing is undertaken in a powered rowing tank with the water flow rate set to simulate a boat speed of 2.3 m/s. In addition to two or three one hour periods of FES ergo-rowing per week, a further one hour per week of FES rowing is undertaken in an adapted station in the turbine powered rowing tank at the London Regatta Centre (McLean, 2002).

20.2.3 Case Report FES Sculling

Subject (6), shown in Figure 20.5, has successfully demonstrated FES sculling. He had two 1 hour sessions on the simulator followed by five 8 minute sessions with 5 minute rest periods in the tank. Initial on-water training began following the second tank session using the Alden 18 double with the coach in the bow seat. After three 30 minute sessions with the coach he began single sculling in the Alden 16, shown in Figure 20.6. He now regularly sculls on-water in typical rowing sessions exceeding 1,000m.

Figure 20.5. The adapted stateroom module shown in Figure 20.3 is shown installed into the Alden 16 shell. Subject (6) is shown dockside with the leg stabiliser in place, testing the control switch fixed to the oar handle.

Figure 20.6. Subject 6 in the adapted Alden 16 shell. The floats mounted below the oarlocks prevent capsize.

20.3 Discussion

We have observed that all the FES rowers progressively increased their strength and endurance at rowing, beginning with distances of a few tens of meters and progressing to many thousands of meters. For example, subject 1 began in June 2004 and was only able to attain a few hundred meters before quadriceps were fatigued. In November 2004, at the BIRC, he had adopted a 30s FES row/30s arms only (seat clamped) split, to allow the quadriceps to recover yet make good time for 2,000m. The split was progressively staged to 40:20 then 50:10 then non-stop for 2,000m. In September 2006, in an officially timed marathon, 50,000m was achieved in 5hrs 49mins. He now routinely trains non-stop at 3-10,000m. Clearly, there have been profound changes in his physical fitness and stimulated muscle endurance. These changes are now the subject of further studies.

FES rowing is offered at ASPIRE (www.aspire.org.uk), the Steadward Centre at the University of Alberta (www.steadwardcentre.org/), and the London Regatta Centre (www.london-regatta-centre.org.uk). Internet rowing (RowPro), (www.digitalrowing.com), is used to link participants. FES rowing offers those with SCI a range of work-out intensities and volumes. For many the fun of FES rowing, and the associated social activity, is enough. A critical mass of participants has built up who share ideas and suggest modifications to the technology and training programmes. The competitive element has ensured continuous improvement in performance and technology.

In the BIRC 2006, medals were awarded based on the average time calculated from an individual's previous 10 attempts at 2,000m just prior to the championships. If this time was beaten by 10s, the rower achieved Gold, less than

10 but over 8s Silver, and below 8s Bronze. In this way the awards were considered fair but competitive across the widely different exercise capacities.

In September 2007 at the World Rowing Championship in Munich, subject 6 achieved Gold in the adaptive fixed seat class AM1x. In training he was able to achieve a higher workout intensity using FES than fixed seat rowing: the increased cardiovascular fitness may have contributed to his success (Hettinga, 2007). The application of FES rowing for cross-training paraplegic athletes is now being investigated.

20.4 Acknowledgements

We thank the Henry Smith Charity, the Trusthouse Charitable Foundation, Inspire UK, the David Tolkien Trust and SCITCS (Northern Alberta) for their support.

20.5 References

Durstine JL, Grandjean PW, Davis PG, Ferguson MA, Alderson NL, DuBose KD (2001) Blood lipid and lipoprotein adaptations to exercise: a quantitative analysis. Sports Medicine, 31(15): 1033–1062

Hettinga D, Andrews BJ (2007) The feasibility of FES rowing for high energy training and sport. Neuromodulation, 10(3): 291–297

Hettinga D (2007) Adaptive rowing. In: Secher NH, Volianitis SS (eds.) Handbook of sports medicine and science: rowing. Blackwell Publishing, UK

Jacobs PL (2004) Pulling shoulder pain away. Sports 'N Spokes, July: 38–42

Manns PJ, Chad KE (1999) Determining the relationship between quality of life, handicap, fitness, and physical activity for persons with SCI. American Journal of Physical Medicine and Rehabilitation, 80(12): 1566–71

McLean (2002) The London Regatta Centre and ARUP Rowing Tank. The ARUP Journal 1: 21–3. Available at: www.curlewrowingclub.co.uk/lrctank.pdf (Accessed on 8 December 2007)

Poulton AS, Andrews BJ (2004) A simple to program but sophisticated distributed control system for surface FES applications. In: Proceedings of the 9th Annual Conference of the International FES Society, Bournemouth, UK

Tanasescu M, Leitzmann MF, Rimm EB, Willett WC, Stampfer MJ, Hu FB (2002) Exercise type and intensity in relation to coronary heart disease in men. JAMA, 288: 1994–2000

Verellen J, Vanlandewijck Y, Andrews BJ, Wheeler GD (2007) Cardiorespiratory responses during arm ergometry, functional electrical stimulation cycling, and two hybrid exercise conditions in spinal cord injured. Disability and Rehabilitation: Assistive Technology, 2(2): 127–132

Wheeler G, Andrews BJ, Davoodi R, Nathan K, Weiss C, Jeon J et al. (2002) Functional electric stimulation assisted rowing. American Journal of Physical Medicine and Rehabilitation, 83: 1093–1099

Chapter 21

Universal Access to Shopping: Apparel Acquisition Preferences for the Working Woman with Physical Disabilities

K.E. Carroll and D.H. Kincade

21.1 Introduction

Consumers with physical disabilities often find the acquisition of suitable ready-to-wear apparel to be problematic and in some cases impossible. The number of female consumers with physical disabilities who are employed is expanding and these consumers need access to affordable, functional and attractive apparel. In spite of multichannel retailing being available, consumers with disabilities tend to prefer to shop in traditional brick and mortar stores. However, significant problems exist in finding the right product, accessing stores and fitting rooms, and finding personnel who are willing to listen and/or have some knowledge of disability. The current qualitative study involving a sample of working women with a variety of physical limitations (n=9) supports the evidence that the shopping preference of consumers with disabilities is the brick and mortar store. Their reasons for this preference included being able to see and touch the fabric, and getting a feel for how the garment looks on the body before making the purchase decision. In addition, price points were examined to see exactly how much money consumers with disabilities were willing to pay for an apparel product (jacket). This study has implications for retailers and apparel manufacturers who wish to address the diverse shopping needs of this often neglected segment of consumers.

Retail shopping, considered one of the *instrumental activities of daily living* (National Center for Health Care Statistics, 2007), presents a multitude of challenges to the consumer with physical limitations. Universal access to retail locations for someone with a physical disability, although desirable, is problematic if not impossible with today's retail environment, and as a product sector, ready-to-wear apparel is often not appropriate in styling or comfort for these consumers. Other shopping options exist, which also have their own unique set of problems, such as custom alterations to existing garments. Finding a service to alter existing apparel can be time-consuming and expensive, especially if mobility is limited

(Braaten, 2000). Some companies manufacture specially designed functional clothes and offer them through catalogues and Internet sites, but these items are often limited in style, colour and size. Buying from these sources seems to be a ready made solution but they do have the added problems of inability to feel and try-on products and complicated systems of payment and returns. In the mid 2000s, consumers with physical disabilities make up a relatively small segment of the retail apparel market (De Klerk and Ampouseh, 2002); however, more people with physical disabilities are being mainstreamed into US society and, with the ageing US population, many more people will have some form of physical disability. With these changes, a need exists to improve the availability of well-fitting, stylish, ready-to-wear apparel to accommodate a group of consumers who have not traditionally been well-served by the apparel industry. Carroll and Kincade (2007) found that working female consumers with disabilities are interested in levels of fashion, quality, modesty, comfort and price and that satisfaction of these criteria goes a long way towards enhancing self-esteem and helping job performance. For retailers and apparel manufacturers to meet the needs of these consumers, additional research is needed.

21.2 Literature Review

The number of potential consumers with physical disabilities looking for easily accessible and affordable apparel has increased since the passage of the Americans with Disabilities Act in 1990. According to US Census Bureau figures, 43% of people with a severe disability aged 21–64 were employed in 2002. In addition, the majority of people with disabilities are female (19.5%) compared to 16.7% of males (Steinmetz, 2002a–c), and, as female consumers, will be more interested in apparel options and shopping availability (Goldsmith and Flynn, 2004).

The ready-to-wear apparel industry has had difficulty addressing the basic needs of these consumers while maintaining profit margins due to (a) the uniqueness of almost every type of physical disability and (b) the difficulties in producing apparel to fit body types that vary from industry standards (Thoren, 1997). However, retail executives interviewed about the future of retailing generally advise that meeting the needs of consumers with physical disabilities is possible, although the focus may be on niche markets backed by strong product development (Yoh and Gaskill, 1999). Thoren (1997) suggests that apparel companies responsible for design, production and distribution of such products need to improve the quality of several aspects of the system, including: (a) provision of sufficient choice of suitable apparel products for a variety of situations, including work, (b) development of a system that will produce affordable and satisfactory apparel for the physically disabled consumer at both a symbolic (psychologically comfortable) and functional (easy to use) level, (c) cost-effectiveness in production and marketing, and (d) communication about available apparel products. Thoren, as well as many others in the apparel and retail industry (Carroll and Kincade, 2007), state that the most common solution to the accessibility problem for persons with physical disabilities is for retail and other businesses to provide services whereby consumers can buy specialised apparel, preferably using mail order or Internet. In Thoren's study, 65 people with varying levels of disability were interviewed. Contrary to industry suggestions, the subjects did not want to use a mail order system. The mail order system and the newer Internet system have a number of problems that all persons experience – problems with returns, inability to feel or try-on apparel, misrepresentation of colours, and payment issues (Endo and Kincade, 2005). In addition, Thoren found that his subjects did not want to use specialised catalogues, which made them feel as if they were being "labelled" with a disability. Carroll and Kincade (2007) found that working female consumers with disabilities did not want to be seen as different or otherwise unable to perform their jobs.

Many researchers have studied the apparel shopping habits of retail consumers with physical disabilities; however, much of the research is several years old and was carried out before the widespread use of the Internet and other retail changes. O'Bannon *et al.* (1988) looked at the type of information sought and the information sources used by wheelchair-bound consumers when purchasing apparel. They found that people with disabilities used a variety of information sources (contrary to Thoren's study) to make their apparel purchase decisions, with family members and mail-order catalogues ranked highest. Changes in ranking of mail-order may have happened with time, as this study is 10 years before Thoren's study. Retail sales personnel were also viewed as sources for making apparel purchase decisions. The most important types of information sought were price, care information, garment coordination, fibre content, budget, and stylistic appropriateness. Fashion trends, in contrast to findings in Carroll and Kincade's (2007) study, were deemed the least important.

Macdonald *et al.* (1994) surveyed consumers with a variety of physical disabilities (n=625) to acquire information on their shopping habits. Department, discount and mass merchandise stores were the most frequented, while specialty and outlet stores and mail order were used less (less than 10% each). Thrift stores, medical supply stores and television (less than 1%) were used very infrequently. Seventy-eight percent preferred stores to be situated in a traditional mall, in spite of the accessibility barriers encountered, such as steps, curbs, fitting rooms and restrooms; staff who were either unhelpful or lacked knowledge about their needs, and lack of products. Eighty-three percent preferred to make their own apparel decisions in a store, demonstrating a desire for independence. When asked about availability of information on sources of apparel for persons with disabilities, 74% were unaware of any companies which specialised in apparel for the disabled. Forty-seven percent were prepared to pay more for functional and fashionable apparel if it were made available. The researchers recommended the formation of a national information centre to disseminate apparel information to this target market and encouraged retailers to do more to take leadership in this area.

De Klerk and Ampousah (2002) found that 59% of women with physical disabilities aged 21–60 (n=40) preferred chain stores over other types of retail establishments. In an observation of 40 shops of varying format – hypermarkets, department stores, chain stores and boutiques, the following problems were commonly encountered: difficult access, lack of parking and/or lifts, out-of-reach display racks, narrow passageways, inaccessible fitting rooms, lack of assistance and little or no facilities for on-the-premises alterations.

Previous research about shopping habits of women with physical disabilities has focused on variations within the brick and mortar format. No previous studies were found comparing shopping in brick and mortar stores with alternative non-store shopping, such as catalogues, television, the Internet and direct selling. In addition no studies were found that compared the shopping habits of working women with physical disabilities with those of working women without physical disabilities.

As alternatives for purchasing mass-produced apparel products, catalogues, television, the Internet and direct selling are often mentioned as practical solutions, However, online retailing, while an established part of the retail world, has not become the cataclysmic industry-changer that analysts predicted in the mid 1990s (Linn, 2007). Many of the problems that consumers have with the Web, such as: product information, measurement information and product availability would no doubt be compounded if the consumer had some type of physical disability. For example, these consumers typically have lower incomes and less access to technology and desire the socialization that shopping in a traditional retail environment provides.

21.3 Method

This exploratory research was completed to investigate the current apparel shopping practices of working women with physical disabilities. Understanding and preparing distribution to a target customer is an essential part of the process of product development for apparel. Carroll and Kincade (2007) outlined a conceptual framework in which distribution was the fifth and final step in a five part process, the only step not to be empirically tested in their study. The current investigation will guide the development of that final step in future research. The following objective was used to investigate the distribution step of this process:

Gather detailed information about current acquisition options from a sample of working women who are affected by a variety of physical disabilities.

A small sample (n=9) of working women with a variety of physical disabilities were selected via snowballing and other purposive techniques (see Table 21.1). Because of US federal legislation to protect the privacy of health information about individuals, lists of persons with disabilities are not available for random sampling techniques. In addition, with limited information about current shopping practices of these individuals (*i.e.* most research is several decades old), an interview format – possible with a small sample – with the ability to probe on some questions was selected as the best way to initiate this exploratory research.

Table 21.1. Participants and their physical disabilities

Participant Number	Disability
01	Fibromyalgia
02	Post-polio syndrome
03	Carpal tunnel syndrome both wrists Osteoarthritis Double hip replacement
04	Spina bifida
05	Fibromyalgia Chronic fatigue
07	Severe trauma from auto accident
08	Peripheral neuropathy Auto-immune disease Vasculitis
09	Rheumatoid arthritis Frozen shoulder
10	Rheumatoid arthritis

Face-to-face interviews regarding current apparel preferences and apparel acquisition practices were conducted. A questionnaire was used to maintain reliability and consistency in the interviews across participants. The questions covered seven options of retail locations and the reasons why these choices were made. In addition, participants were asked to make a choice of price that they would be willing to pay for an item such as a loose-fitting jacket that they could wear to work, if was comfortable and aesthetically pleasing. The validity of the price information gathered in traditional surveys is often questioned; however, these purposely selected participants were willing to discuss price. Determining the price that consumers would be willing to pay for this type of product would help manufacturer and retailer set a target price point for production and sale.

21.4 Results

21.4.1 Acquisition Choices

The objective stated that the researcher would gather information about current acquisition options from the sample of working women with physical disabilities. Table 21.2 shows the choices made by participants when asked to identify preferred methods of purchasing the suggested garment if it were available in the ready-to-wear market. Six options were given in the interview question, which encompassed a variety of acquisition methods, and a seventh option allowed for other options not thought of by the researcher. Participants were allowed to select as many options as they wished and could explain their choice(s), a benefit of the small sample and interview format. Frequencies are provided in Table 21.2 and further explanations of these choices, as provided by the participants, are given in the following sections.

Table 21.2. Acquisition choices made by participants

Current purchasing practice	Frequency
Traditional brick and mortar store	9
Online Shopping	0
Catalogue Shopping	5
TV Shopping	0
Direct selling	5
Custom designed	4
Other	0

Traditional brick and mortar shopping was selected by all nine Participants. Reasons stated for preferring this over other methods were varied. They included; more availability of various styles, the option to feel fabric and the possibility of trying on a garment. "The feel is the first thing that hits me. I'll buy something that may not necessarily look trendy but it feels good. I'll try on the apparel and feel how it performs while sitting, standing, stretching, bending and walking (Participant 05)."

Disadvantages of the traditional brick and mortar store were reported as being a lack of facilities for people in wheelchairs and on crutches, and sales assistants being ignorant about people with physical disabilities. This echoes the findings of DeKlerk and Ampouseh (2002). One Participant related the following experience:

> I had a friend with me and I borrowed a transportation service wheelchair. This sales assistant kept bringing up dresses and saying to my friend 'do you think she's going to like these?' I finally got angry with her and told her that I was the one buying the dresses so she should talk to me, not my friend."
>
> (Participant 04)

Another problem with brick and mortar stores was the inaccessibility of fitting rooms. "The accessibility in some of these stores in some of the dressing rooms leaves something to be desired...I take time in dressing rooms – sitting, standing, stretching, bending – I really want to try and get a feel of how it will feel once I've worn it a few hours" (Participant 05). This type of problem has been evident in other studies involving consumers with disabilities (MacDonald et al., 1994).

With online shopping, privacy issues, such as giving out a credit card number over the Internet or over the telephone were mentioned as factors that prevented Participants from choosing this acquisition channel. Participants were also not enthusiastic about an acquisition method which did not allow them to see the product first-hand before a purchase decision was made. These findings reflected common complaints about Internet shopping (Reichheld *et al.*, 2000; Miyazaki and Fernandez, 2001).

Catalogue shopping as a method of acquiring apparel was acceptable for five participants. Advantages of buying from a catalogue, as opposed to other methods, included availability of a toll-free number to call for ordering and for customer service, ease of returning products, availability of good quality specialty apparel in special sizes, ability to plan a wardrobe over a long period of time before placing an order, and reliability in sizing and apparel quality. Disadvantages included variation in fit, lack of aesthetically pleasing apparel, variation in quality, and stockouts. "There are some catalogues whose quality I trust and whose sizes I'm familiar with. With Land's End, I know what I'm getting in the sizes though I don't like their styles. Other catalogues, I like their styles but I don't know what I'm getting" (participant 08). Although some participants were comfortable with using catalogues for apparel shopping, many also wanted to be able to handle garments. "I look through catalogues but hardly ever buy anything because touching and feeling the garment is so important to me" (participant 10). Catalogues are an important source of information about products for people with physical disabilities and have been recommended as a purchasing method by other researchers (Koester and Leber, 1984), but they remain limited in appeal for these consumers.

Direct selling was popular with five participants. Reasons given for choosing this method included privacy, meeting other people with similar problems, product-educated salespeople, the ability to try something on without feeling pressured to buy, and perceived longer decision-making time than other methods. Disadvantages associated with this method included a dislike of getting together with other women to buy apparel and feeling pressured to buy. These contrasting ideas of the advantages and disadvantages might suggest that participants are not very familiar with this method of acquisition and that more education about this method might encourage more people to consider it as an acquisition option.

Custom design was chosen by four participants, mainly based on their personal experiences. Personal experiences with customised apparel, by the participants, included living in other countries where custom design and production is expected and inexpensive for professional apparel, having an arrangement with a manufacturer to design custom pieces, and having mothers and grandmothers who had always constructed apparel for them. Participant (05) who had international experiences was concerned about using similar methods in this country. She said, "then again you can get wonderful results too. I haven't tried custom work in this country, it would seem so expensive to me." The primary reason participants did not choose custom design as an acquisition option was the lack of availability of affordable custom dressmakers in their area. Participant 09 stated, "I would probably have it custom designed if I could find someone."

21.4.2 Price Range

Participants gave varied responses to the question of how much they would expect to pay for a garment such as a jacket that could be worn to work (see Table 21.3). Participant 04, who chose the lowest price category, was very conscious about the lack of spending power for people with severe physical disabilities. She said, "[the price] would depend on how much money I have. Right now under $25 would be right, but please remember that only about 31% of people with disabilities have jobs, which puts them into the lowest [income] category. You do have to think about that." Five Participants chose a potential price between $26 and $75 because they typically spent that much on a jacket for work and related the prototype closely to this type of garment. Participant 05 stated that she never spends more than $50 on a jacket and always shops the sales, which limits her choice of styling. Participant 10 was accustomed to paying more than $50 because she was accustomed to buying large sizes, and assumed larger sizes meant higher prices.

Table 21.3. Price range preferred by participants

Price range preferred	Frequency
$0–$25	1
$26–$50	3
$51–$75	2
$76–$100	0
>$100	3

Three participants stated a willingness to pay more than $100 for the product, including Participant 07 who would be "happy to find a garment that she liked, could wear comfortably, and which looked good" on her. "If [I could find] a sapphire blue, flowing kind of thing and it had a $150 price tag on it, I would pay it." Participant 02 equated price with workmanship and how the proposed jacket would be manufactured. She said "That's something I would have to think about. If something really appeals to me and fits me I don't mind paying a good price. If it was made of a good fabric with good workmanship I would easily pay between $100 – $125 for [a jacket]".

Comments from the participants on pricing were dependent upon type of garment, function of garment, colour and fabrication of garment, and the construction or quality of the garment. These results support Feather's (1976) findings that people with disabilities are willing to pay more because they are used to not finding any clothes that accommodate their needs.

21.5 Conclusions

Access to functional and attractive apparel is desired by but difficult to achieve for working women with physical disabilities. Shopping is considered an instrumental activity of daily living, yet many people with disabilities find problems in getting to stores, accessing them and finding suitable products. Results from these participants confirmed previous studies in that brick and mortar stores are overwhelmingly the most popular place for working women with physical disabilities to shop for apparel. They chose this retail format because they wanted the range of choice and the ability to touch and feel the product. However, these types of stores have major disadvantages such as inadequate facilities and ineffective salespeople. The study also revealed that catalogues, home demonstration and custom design are distribution choices for working women with physical disabilities, but some women continue to see these as choices with disadvantages and undesirable features. Online shopping and television were not chosen by any subjects in this study.

The prices the target customers would be willing to pay for a satisfactory garment ranged from below $25.00 to over $100.00. Some of these participants were willing to pay more for a product they knew was "right" for them, but the perceived price of such as garment might be a deterrent considering the spending power of people with disabilities. Income level of this group of working women was not assessed in this study, but the positions of the participants ranged from part-time administrative assistant to college professor. More work should be done to segment income level in future studies.

Obviously, with such a small sample of working women with disabilities, no firm conclusions can be drawn from the current study. However, the results tend to reinforce evidence that brick and mortar is the preferred apparel shopping method for these consumers. In addition, it appears, from these preliminary findings, that the industry needs more easily accessible facilities, stores that are easy to navigate through, effective and empathetic salespeople, and quality products. A much larger and more detailed study should now move forward with this information, to bring enough numbers to the table in order to convince retailers and manufacturers that changes should be made. In this larger study, detailed demographic information and qualitative responses should be obtained to discover much more about these consumers and their apparel acquisition preferences. Further research should also be carried out on the potential for home sales and catalogue shopping for these target customers.

In many ways solving the problem of availability of apparel and shopping accessibility, for working women with physical disabilities, would result in a win-win situation. These consumers need the right product distributed to them with the right approach, while the apparel industry needs to establish evolving niches with new target customers.

21.6 References

Braaten E (2000) Personal communication. 28 November 2000
Carroll KE, Kincade DH (2007) Inclusive design in apparel product development for working women with physical disabilities. Family and Consumer Sciences Research Journal, 6(35): 289–315

De Klerk HM, Ampousah L (2002) The physically disabled South African female consumer's problems in purchasing apparel. International Journal of Consumer Studies, 26(2): 93–101

Endo S, Kincade DH (2005) The developing direct relationship between a manufacturer and consumers: four models of interaction. Journal of Fashion Marketing and Management, 9(3): 270–282

Feather BL (1976) The relationship between the self-concept and apparel attitudes of physically handicapped and able-bodied men and women. Unpublished PhD-thesis, University of Missouri, Columbia, MO, US

Goldsmith R, Flynn L (2004) Psychological and behavioral drivers of online clothing purchase. Journal of Fashion Marketing and Management, 8(1): 84–95

Koester AW, Leber DA (1984) Diffusion of information about apparel to orthopedically disabled adults. Home Economics Research Journal, 13(2): 153–158

Linn A (2007) Online shopping growth to slow in next decade. Available at: www.msnbc.msn.com/id/20321999/ (Accessed on 14 October 2007)

MacDonald NM, Majumder RK, Bua-Iam P (1994) Apparel acquisition for consumers with disabilities: purchasing practices and barriers to shopping. Apparel and Textiles Research Journal, 12(2): 38–45

Miyazaki AD, Fernandez A (2001) Consumer perceptions of privacy and security risks for online shopping. The Journal of Consumer Affairs, 3(1): 27–44

National Center for Health Care Statistics (2007) Instrumental activities of daily living. Available at: www.cdc.gov/nchs/datawh/nchsdefs/iadl.htm (Accessed on 1 September 2007)

O'Bannon PB, Feather BL, Vann JW, Dillard BG (1988) Perceived risk and information sources used by wheelchair-bound consumers in apparel purchase decisions. Apparel and Textiles Research Journal, 7(1): 15–22

Reichheld FF, Markey Jr. RG, Hopton C (2000) E-customer loyalty: applying the traditional rules of business for online success. European Business Journal, 12(4): 173–179

Steinmetz E (2002a) Americans with disabilities: 2002. Available at: www.census.gov/hhes/www/disability/sipp/disab02/awd02.html (Accessed on 1 September 2007)

Steinmetz E (2002b) Household economic studies. Available at: www.census.gov/hhes/www/disability/sipp/disab02/awd02.html (Accessed on 1 September 2007)

Steinmetz E (2002c) Current population reports P70-107. Available at: www.census.gov/hhes/www/disability/sipp/disab02/awd02.html (Accessed on 1 September 2007)

Thoren M (1997) A new approach to apparel for disabled users. In: Kumar S (ed.) Perspectives in rehabilitation ergonomics. Taylor and Francis Limited, London, UK

Yoh E, Gaskill LR (1999) U.S. retail executives' perspectives on the future of retailing. Journal of Fashion Marketing and Management, 3(4): 324–336

Part V

Inclusive Environments

Chapter 22

Is Remodelled Extra Care Housing in England an Inclusive and 'Care-neutral' Solution?

R.E. Mayagoitia, E. van Boxstael, H. Wojgani,
F. Wright, A. Tinker and J. Hanson

22.1 Introduction

This paper reports findings from a two-year EPSRC-funded study that examined how sheltered housing and residential care homes in England had been remodelled to Extra Care Housing (ECH). ECH is a relatively new type of housing for older people, which aims to provide flexible care while fostering independence, though no agreed definition of ECH exists. Ten case studies of social housing schemes that had been remodelled into ECH were chosen from different regions of England. Access and assistive technology in flats and communal areas were audited by a multidisciplinary team. The aim of this paper is to examine the impact that some of the successes and failures in improving accessibility during remodelling had on care provision. Even after remodelling, the design and layout of most buildings did not fully satisfy current accessibility standards, leading to an increased need for care for some tenants once the building was reoccupied. Successful examples of accessibility, assistive technology and care integration required both active tenant involvement and creative design input from care staff, architects and builders who were assistive technology and accessibility aware. It will be argued that for new and remodelled ECH buildings to be care-neutral, designers need to work towards the most inclusive model of ECH.

It is generally accepted, and indeed encouraged by public policy, that it is best for older people to stay in their family homes for as long as possible (Royal Commission on Long Term Care, 1999). This is often achieved by a combination of home care assistance, home modifications and assistive technology (AT). However, when older people do choose, or need, to move, there are a number of options on offer (Figure 22.1). Making the wrong housing choice in later life could oblige someone to move house every few years or so in old age, when it can be a particularly traumatic experience (De Coninck, 2004). To avoid this predicament, housing solutions like ageing in place (WHO, 2004) and lifetime homes (Carroll *et al.*, 1999) have been developed to promote the design and building of dwellings that will suit people from birth to death, regardless of their changing needs and levels of ability.

One alternative to these 'mainstream' approaches to housing in later life is Extra Care Housing (ECH), a relatively new type of age-specific, grouped housing for older people that aims to provide flexible care while fostering independence in a self-contained dwelling. This may sound slightly paradoxical in that both care and independence are a presupposed part of ECH, but in an ideal situation this model of housing and care is believed to lead to a 'win-win' situation where, through carefully tailored support packages, older people both receive as much care as they require and achieve as much independence as they aspire to.

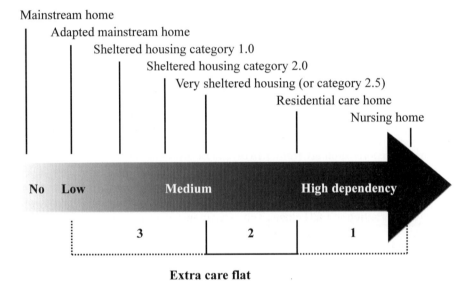

Extra care flat

Figure 22.1. Above the arrow are the housing options covering progression from total independence to total care in the UK. Below the arrow are the possible locations of ECH, according to the three models of provision described by the Care Services Improvement Partnership (2006).

By contrast with sheltered housing, which was very clearly defined (Ministry of Housing and Local Government, 1969), there is no single agreed definition as to what constitutes ECH (Tinker *et al.*, 2007). However, the Care Services Improvement Partnership (2006) discusses three models of ECH that are in use in the UK at present. Model one is known as "An alternative to care homes" and in Figure 22.1 (above) it would only encompass block one, demarcated by the dotted line to the far right, below the arrow, for high dependency but not nursing tenants. Model two known as "Ageing in place" would encompass the middle (solid line) and right hand (dotted) blocks (1 and 2), starting from medium dependency. Finally, Model three entitled "A home for life" (also in Figure 22.2) would encompass all three blocks (from the left hand dotted onwards), from low dependency (or maybe even no dependency) to high dependency tenants; a place from where the majority of older people would not need to move once they had relinquished their family home. Whichever model is adopted, one objective of extra care housing is to bridge the gap between living in a mainstream home with little or no support and living in residential or nursing care with almost no independence.

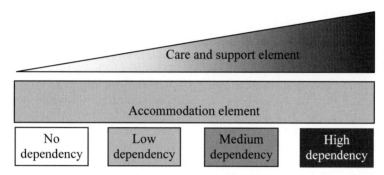

Figure 22.2. A "Home for life" model of ECH. This model assumes the building is the base for provision and is fully accessible and inclusive, and flexible enough to accommodate all tenants. The care and support element is assumed to be adaptable as tenants' needs change, either short-term (*e.g.* recovery from illness) or long-term (*e.g.* with increased frailty). Diagram redrawn from Care Services Improvement Partnership (2006).

However, during the course of this remodelling project it soon became apparent that in a number of cases the failure to achieve a fully accessible building, when it was converted from sheltered housing or residential care to extra care, subsequently required an increase in care provision to compensate for the inadequacies of the building. Far from fostering independence, the poor design of a building or its fixtures and fittings was producing 'architectural disability' in its occupants (Hanson, 2001). As a result of this, the concept of a building being 'care-neutral' was born.

The basic premise is that a building can be regarded as care-neutral if it does not have any impact – either positive or negative - on the support and care regime that the building supports. Defined in terms of care, the building's residents would not require the help of a caregiver to overcome the building's physical or other shortcomings. A building can be regarded as care-positive if its inclusive design allows or even encourages the independence of its tenants; a building is care-negative if its poor, inaccessible design has adverse quality of life ramifications for the tenants or increased cost implications for the service provider. If additional care is needed to overcome architectural barriers, this care will need to be paid for. If tenants are less independent due to the built environment, their quality of life suffers. The concept is applicable to new as well as remodelled buildings.

22.2 Materials and Methods

Ten social housing schemes, in the different regions of England, remodelled into ECH from either residential care homes (two cases) or sheltered accommodation (eight cases) since 2000 were chosen. The changes resulting from the remodelling to the building, care, access and AT, and the views of staff and tenants as well as those of the architects, contractors and quantity surveyors were obtained through site visits and interviews. The changes to the schemes were costed.

Within each scheme, a range of typical flats was visited to audit their accessibility and AT. These flats were chosen because of their design features, as represented in the building plans received in advance. The scheme manager contacted the tenants at the chosen flat types. Those who agreed to be visited formed the sample. An architect and an occupational therapist carried out the access and AT audit while the rehabilitation engineer usually stayed with the tenants in the sitting room and asked them about how they found the housing scheme and about their AT use. The communal areas of the buildings were also audited for AT and accessibility. In total 44 flats were visited; 40 flats were occupied, 4 tenants had partners, 37 tenants were in extra care housing and 3 tenants were in flats designated as 'sheltered' that were part of a housing scheme that also contained dwellings designated as 'extra care'.

22.3 Results

Table 22.1 (below) contains the main elements found in the 10 ECH schemes. Some of the elements such as community alarms and laundry facilities were found to be present in all or most of the case study schemes, while others such as consulting rooms or internet access were under-represented.

A total of 356 AT devices of 52 types were found in the flats. Only six types were classed as exclusively linked to the provision of care and were: community alarm (44), profiling bed (13), shower chair (10), mobile hoist (3), ceiling hoist (3) and transfer pole (2). As a small aside, wheelchairs were classified separately because they can be propelled independently as well as by a caregiver and all but two of the chairs that were found in the flats were self-propelled.

Being remodelled rather than purpose-built, the buildings did not fully meet current accessibility standards (The Stationery Office, 2004) and what each scheme provided was different from the others. This is why wheelchair accessibility is described as patchy in Table 22.1. In the case of some corridors and lifts, there simply was not enough space to widen them and no viable alternative could be found within the site. However, there were a number of accessibility breaches that could have been corrected as part of the remodelling process but were not, including: heights of sockets and switches; the position of radiator valves, alarm pull cords and intercoms; stiffness of fire and front doors, and heights of bathroom and kitchen fixtures. A number of examples were found of how poor accessibility had a negative impact by requiring increased care. Two have been chosen here for illustrative purposes.

Table 22.1. Elements distinctive of Extra Care Housing found in the 10 schemes studied

Element	Total	Element	Total
Flat's own front door	10	Staff room	8
Social alarm	10	Hair dressing facilities	8
Wheelchair accessibility (patchy)	10	Communal dinning room	7
Communal lounge	10	Guest room	7
Flexible care	10	Commercial kitchen	7
Laundry room (shared with staff 6)	10	Staff WC	7
Manager's office	10	Provision of communal meals	6
Carer's office	10	Waking night staff on site	5
Communal activities	9	Sleeping night staff on site	4
Staff sleep-over area	9	Consulting room	4
Assisted bathroom	9	Tenant's shop	2
Scooter/wheelchair store	9	Internet (dial-up)	2
Tea kitchen	8	Manager's flat	2

22.3.1 Example 1: Lifts

Lifts are the first example. Several of the tenants visited reported having difficulty operating the lifts. They found that the lift was located too far away from their flat, the buttons were difficult to reach or see, the space inside the lift was tight, the shutting time of the door too quick and the movement of the lift too jittery. Some were frankly scared of using the lift. This meant that they either gave up on using the lift altogether or needed a neighbour or member of staff to accompany them when moving around the housing scheme. For some, this in turn reduced their opportunities for social interaction within the housing scheme. For others it resulted in their having to depend on a carer to collect them for the communal meal and return them to their flats some time afterwards.

Most schemes had only one lift and, if it broke down, tenants were stranded. Lifts that barely complied with wheelchair standard dimensions were too small to accommodate people riding the scooters that so many older people like to use for shopping. Only two lifts were found that were large enough to fit a stretcher but neither of them was large enough also to accommodate an attendant travelling with a person on the stretcher. The impracticality of having to take a person who is ill down the stairs on a stretcher was mentioned by both tenants and caregivers.

22.3.2 Example 2: Baths and Assisted Bathrooms

In one scheme, the baths in the flats on the first floor were left during the remodelling, under the pretext of offering tenants choice (though running out of money was also mentioned), while the flats on the ground floor were given new level-access showers (Figure 22.3). Most of the tenants on the floor with baths, unlike their counterparts below, either needed a caregiver to assist them in bathing or used the communal shower located down the corridor, instead of the baths in their own flats, on a daily basis. Leaving the old baths in situ therefore resulted in additional care costs and defeated the ECH objective of having a self-contained home, encouraging independence.

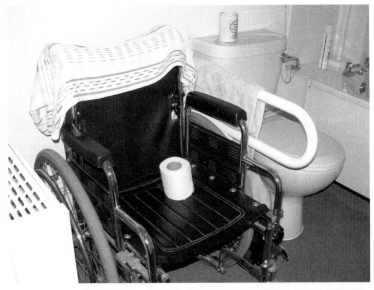

Figure 22.3. Bathroom of a wheelchair user where the bath (far right) was left during remodelling

Of course, not all baths are inaccessible. In one scheme new walk-in baths were installed in every flat, at great expense, during the remodelling. One by one the tenants asked for the walk-in baths to be replaced by a level access shower. There is only one walk-in bath still in place in this particular building. The main reason for rejection was that during the long time it took for these baths to fill and empty, the tenants grew very cold. Tenant input or "try before you buy", which was not sought, could have been very valuable when making the decision to install these baths during remodelling.

Assisted bathrooms in the remodelled schemes usually had a very institutional (as opposed to homey) look and feel, probably because of the presence of a bath that integrated a lift, or a hoist in the room. In one scheme an effort had been made to make it look more like a commercial spa, with modern decor and additional curtains for increased privacy. Three of the tenants visited (7.5%) reported using the assisted bath in their respective schemes, with the intended carer assistance, about once a week.

22.3.3 Example 3: Communal Laundries

The next example has been chosen to illustrate how accessibility had a positive impact on inclusivity. In one scheme a laundry was found to have washing machines and driers placed on plinths, with front loading and easy to use and see controls on the front panel. The room had good illumination and ventilation and it had ample circulation and turning space. The sink for hand washing had lever action mixer taps. One of the tenants proudly reported that the laundry was "so good because we designed it".

There were two tenants' laundries in the building and this one had recently been refurbished (Figure 22.4). The tenants had been consulted before its refurbishment and had rearranged the space and chosen the machines to go in it to best suit their needs to do their laundry independently. The other laundry in the building was much less accessible (Figure 22.5). The refurbished laundry is used independently by many more tenants than what the tenants call "the old laundry". This may be allowing some tenants to still do their own laundry rather than depending on a caregiver for this task, though proving a negative is notoriously difficult.

Figure 22.4. Laundry in the same building that has not benefited from a tenant-centred refurbishment

Figure 22.5. Accessible laundry in an extra care housing scheme redesigned with tenant input

22.4 Discussion

Though the concept of inclusivity encompasses far more than accessibility, bringing together the findings of all case studies it can be said that the greatest barriers to inclusion that were found all related to poor accessibility, while the greatest successes were related to taking a user-centred approach together with providing the necessary accessibility. Though only three examples are given here, all the case studies pointed to an inclusive ECH building being also a care-neutral ECH building, with the concept of inclusion encompassing them both.

The accessibility breaches found may be explained but not excused, partly because for most architects and builders these remodelling projects were their first experience of ECH, though most did have previous experience of either sheltered housing or residential care. In many cases poor accessibility had to be compensated for by increased work by care staff, an effect that rendered the building care-negative.

During the course of the project, the "Home for life" model (Figure 22.2) was probably closest to the tacit assumption that the research team made while trying to keep an open mind about what the data would reveal to be the understanding, in practice, of ECH. However the phrase "ECH, also known as very sheltered housing," was also often used by one of our more experienced team members. According to the 1969 definition, the characteristics of 'very sheltered' housing included: a purpose-built block in which both the individual flats and the communal areas were wheelchair accessible; the provision of a residents' lounge, communal laundry, assisted bathroom, and a lift in schemes of more than one story. Services included 24-hour warden /care staff cover, possibly a nurse (permanently) on site, daily mid-day meal provision, help with personal care and an emergency alarm system linked to the warden. Therefore, the team was probably expecting to find all such elements present in all the case studies, but did not (Table 22.1). Nevertheless, the team's aspiration for remodelled ECH remains the most ample model of ECH "A home for life", championing inclusion, encouraging social interaction and making buildings at best care-positive or at least care-neutral because this is the most sustainable long-term solution.

Designers of the remodelling projects may have been confused as to which model of ECH they were working to. They may have consciously or subconsciously supposed that caregivers would be there to compensate for accessibility shortcomings. However, compensation by human helpers is an unsustainable solution to providing housing for older people because it is not cost effective over the design life of the remodelled scheme. ECH urgently needs to be carefully defined, but even beyond that, the significance of inclusion needs to be much better understood by those involved in remodelling social housing into ECH who are looking to fully achieve the twin aims of ECH of providing flexible care while fostering independence, for life.

22.5 Acknowledgments

This research is funded by the Engineering and Physical Sciences Research Council (EPSRC), grant EP/C532945/1. A special thank you to Alan Holmans who did the cost analysis and to all the people who agreed to be visited for this project.

22.6 References

Care Services Improvement Partnership (2006) The extra care housing toolkit www.cat.csip.org.uk/_library/docs/Housing/Toolkit/ECH_Toolkit_Complete_Document_Including_Tools.pdf (Accessed in December 2007)

Carroll C, Cowans J, Darton D (eds.) (1999) Meeting part m and designing lifetime homes. Joseph Rowntree Foundation, York, UK

De Coninck L (2004) De 'International classification of functioning, disability and health' (ICF), een referentiekader om de 'fraile oudere' te beschrijven. Acta Ergotherapeutica Belgica, 2: 8–13

Hanson J (2001) From sheltered housing to lifetime homes: an inclusive approach to housing. In: Winters S (ed.) Lifetime housing in Europe. Katholieke Unversiteit Leuven, Leuven, Belgium

Ministry of Housing and Local Government (1969) Housing standards and costs: accommodation specially designed for old people. Circular 82/69, Her Majesty's Stationery Office (HMSO), London, UK

Royal Commission on Long Term Care (1999) With respect to old age: long term care – rights and responsibilities. The Stationery Office, London, UK

The Stationery Office (2004) Approved document m – access to and use of buildings. The Stationery Office, London, UK

Tinker A, Zeilig H, Wright F, Hanson J, Mayagoitia RE, Wojgani H (2007) Extra care housing: a concept without a consensus. Quality in Ageing, 8(4)

WHO (2004) Ageing and health technical report. Volume 5: a glossary of terms for community health care and services for older persons. Centre for Health Development, World Health Organization, Kobe, Japan

Chapter 23

Designing for an Ageing Population: Residential Preferences of the Turkish Older People to Age in Place

Y. Afacan

23.1 Introduction

In the 21st century, there has been an increase in the ageing population and people with disabilities in the majority of the world. World Health Organization (WHO) estimates suggest that the world total will be more than one billion people aged 60 or over by the year 2025 (Marshall *et al.*, 2004). Thus, ageing population has been a growing concern in Turkey as in other countries. Especially in Turkey, where most Turks live in high-rise apartments, it is important to design housing universally. Turkish culture is based on a close relationship between the older people and their families. Today, rather than preferring specialised institutions, most of the older people want to live in their own houses, where they lived when they were younger. There is a crucial need for alternative daily living environments based on universal design that provides a higher level of accessibility, usability and adaptability for all users regardless of their size, age or ability. This study aims for a cultural understanding of the views of the Turkish older people on designing their homes for ageing in place. It is based on data from empirical research on universal design. The cultural differences between Turkey and places like the US, the UK and Europe, where most of the research in this area has been done, could provide fascinating insights.

Reviewing the literature indicated that there were many debates on ageing taking place amongst assisted living providers, designers, researchers and policy-makers. "When older people become frail, the home environment needs to be more supportive to compensate for their limitations or disabilities" (Pynoos, 1992). The independence of the older people within the boundaries of their homes is prevented by physical, social or attitudinal problems (Kort *et al.*, 1998). Imamoglu and Imamoglu (1992) stated that "in Turkey, examination of current living environments and special housing needs of the older people has been neglected by social scientists, as well as designers, planners and decision makers. In contrast to developed countries, in Turkey old age is not yet regarded as a problem; however it is already becoming difficult to continue existing patterns...". Exploring the changes in life situations of the Turkish older people and their attitudes toward design features for their later years is essential to a scientific study of ageing.

Demirbilek and Demirkan's study (2004) is important in terms of understanding the older people's requirements by involving them in the design process and collecting data by means of participatory design sessions to explore how objects, environments and equipment should be designed to allow for ageing in place. Wagnild (2001), in his most recent survey with 1,775 people aged between 55 and 93 years, stated that although there are barriers to achieving ageing in place, the older people overwhelmingly prefer to grow old where they are. The critical design issue is what can be done to realise this preferred future. In recent years there have been lots of universal design applications in home environments (Ostroff, 1989, 2001; Mueller, 1997; Imrie and Hall, 2001) that aim to investigate the social and physical aspects of daily living environments of older people and to develop design solutions for them. Furthermore, in many countries there are established centres and associations for ageing in place and various technological attempts applied to ageing-related areas, such as housing, personal mobility and transportation, communication, health, work, and recreation and self-fulfilment (Story, 1998; Fozard et al., 2000; Dewsbury et al., 2003; Iwarsson and Stahl, 2003).

However, the design of universal housing is still in its infancy in Turkey. Accommodating ageing in place here is a highly difficult and challenging design task. Although there are cultural and cross-cultural studies on the social psychological aspects of the Turkish older people (Imamoglu and Kılıc, 1999; Imamoglu and Imamoglu, 1992), they do not deal with a universal design approach. This study considers ageing and accommodating the older people as social design problems that cannot be solved by individuals themselves. Architectural solutions informed by ethical concerns, universal design principles and the use of technology can help to overcome these problems. According to Fozard et al. (2000) building technology, architectural knowledge, and smart technology for heating, lighting, and other environmental factors are significant resources for universal solutions in new construction. Designers, providers and users of assisted living should be aware of new technologies to increase the usability of living environments. This study also considers and compares the daily living requirements of the older people living at home with those in an assisted living institution in Turkey, which provides its patients with a universal housing environment combined with emergency help, assistance with hearing and visual impairment, prevention and detection of falls, temperature monitoring, automatic lighting, intruder alarms and reminder systems announcing upcoming appointments and events. Working closely with the older people and being informed of their diverse needs is crucial to the development of enabling design (Coleman and Pullinger, 1993). In this context, this research contributes to the literature by exploring universal design as a critical approach within a cultural perspective and by investigating daily living preferences of the Turkish older people for ageing in place.

23.2 Methodology

23.2.1 Sample

The sample consisted of 48 respondents aged between 76 and 96 years, with a mean age of 83.96 in the home sample and 78.81 in the institution sample. It was selected according to age, sex and dwelling-standards considerations. The idea in the sampling is to present the older people aged 75 years and above, who may have more disability problems and spatial difficulties than those between 60 and 75 years. 32 of 48 (66.6%) respondents, 17 female and 15 male, were selected from the Turkish older people living in their own homes. 16 of 48 (33.3%) respondents, nine female and seven male, were selected from a high quality assisted living environment, the 75th Year Rest and Nursing Home, in Ankara, which was established by the Turkish Republic Pension Fund. There are 168 one-bedroom units of 35m², 47 two bedroom units of 45m² and 18 units of 34m² with special features for disabled people. There are two reasons behind the selection of this institution; the home-like character of the units and the universal design principles behind its features. Each of the architectural features is designed inclusively to increase equal accessibility, privacy, security, safety and usability of spaces within the institution regarding age-related disabilities. The offered services and facilities accommodate a wide range of individual preferences in order to be consistent with the expectations of the older people. Different modes of information presentation, such as pictorial, verbal, tactile and audio-visual, are also used to eliminate unnecessary complexity. Appropriate size and space are provided for approach, reach and manipulation so that hazards, errors and high physical effort can be minimised.

23.2.2 Procedure

A structured interview with 15 questions was conducted with 48 respondents to collect the data. It was held in the respondents' own living environments. The interview questions of the home sample were the same as the questions of the institution sample to provide a valid comparison. The questions were structured to assist the respondents as follows: first, participants were asked to state the characteristics of their living environment; then, they identified the physical barriers within their current or previous living environment and their spatial problems room by room; finally, they were asked about their desired universal design features. The interview was guided by the author in order to elicit responses more comprehensively and later, to generate an in-depth discussion. The results were analysed both qualitatively and quantitatively. In addition to the interviews, photographs were taken to support the verbal responses. Each interview was also recorded on a tape-recorder.

23.3 Results

The recorded interviews and discussions ran for 150 hours. The study systematically analysed the data with SPSS software by means of statistical analyses, such as frequency distributions, cross tabulations and a chi-square test. It examined the findings in detail under the following sub-sections.

23.3.1 Home Environment

Factors Pushing the Home-respondents to Age in Place
The interview results revealed that all 32 respondents identified their current home as the place where they preferred to age although they reported spatial problems and physical barriers within these environments. This preference accorded with Mace's (1988) remark: "The overwhelming preference of older persons is to remain in their homes as they grow older". The main reasons behind this preference in the study were the subjects' memories associated with home, their sense of achievement, their deep attachment to their homes, the cost of living and their fear of change. These factors pushing the respondents to stay in their current homes have a statistically significant relationship to the respondents' relationship status of living alone or living with a spouse or children (x^2=38.882, df=8, α=0, 001, two-tailed). The interview results revealed that nine of 32 respondents lived alone; sixteen lived with their spouse; and seven lived with their children. All of the seven respondents living with their children fear changing their current living environment, because they are adapted to the circumstances and lack a sense of achievement in staying independent and managing unaided. However, eight of sixteen respondents living with their spouses have a sense of achievement in being capable of ageing in their own homes although they are lonely. Moreover, length of stay is another important factor in the preference of the older people for ageing in place. There is a statistically significant relationship between the respondents' living time in homes and the factors pushing them to stay (x^2=24.885, df=8, α=0, 01, two-tailed). Thirteen of 32 respondents, who have lived more than 31 years in their current home, are more anxious to age in place than the other 19.

Spatial Problems Confronted
Considering the spatial difficulties confronted and the physical barriers within the home environments: 32 respondents reported spatial problems caused by long hallways requiring high physical effort, narrow hallways interfering with the manoeuvring space and stair problems such as open risers, narrow treads and landings and inappropriate dimensions. The usability problems caused by long hallways are shared by all 17 of 32 respondents who live in houses bigger than 130m². The long and narrow hallways require much physical effort, maximise repetitive daily actions, obstruct easy access to rooms and cannot meet the changing mobility needs of the Turkish older people (see Figure 23.1a). On the other hand, ten of 32 respondents, who live in smaller houses than 100m², have difficulties in managing everyday life and the problems of ageing due to the

inadequate size and space for approach, reach, manipulation and use (see Figure 23.1b). The narrow hallways do not provide enough space for moving furniture or for manoeuvring, which causes hazards and the adverse consequences of accidental actions. This study analysed the relationship between these spatial problems and respondents' house size (Table 23.1). There is a statistically significant relationship between the house size (in terms of square metres) and the spatial problems (x^2=37.378, df=6, α= 0, 01, two-tailed).

(a)   (b)

Figure 23.1. (a) A long hallway from a 190m^2 house, (b) an inaccessible storage room from a 95m^2 house (by the author)

Table 23.1. Cross tabulation for "house size" and "spatial problems"

	HOUSE SIZE			
SPATIAL PROBLEMS	**80–100 m^2**	**101–130 m^2**	**131+m^2**	**Total**
Long hallways	0	5	9	14
Narrow hallways	7	0	0	7
Stair problems	0	0	8	8
Not enough clear spaces	3	0	0	3
Total	10	5	17	32

In addition to the spatial problems, the respondents were also asked to identify their accessibility and usability problems in day-to-day activities. Figure 23.2 illustrates the inaccessible and unusable features of the three house sizes. These analyses were significant for designers in terms of exploring the relationship between the older people's requirements and universal design performance, identifying the most important features and setting priorities among them, because Wilder Research Center (2002) stated that reconstructing universal design features within a living environment depends mainly on the current home structure and layout.

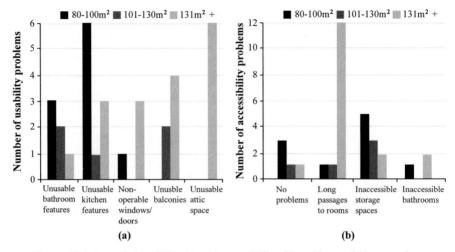

Figure 23.2. Recorded usability (a) and accessibility (b) problems of 32 respondents

The Desired Universal Design Features

Having identified the spatial problems, the respondents were asked to express their desired universal design features. Some examples of the questions are as follows: What do you suggest to improve the unusable kitchen features? What do you suggest to efficiently use the storage spaces? What do you suggest to improve the unusable bathroom features? While answering these questions, respondents gave their ideas about design suggestions. Electrically operated counter tops and pull-out work boards were the most common answers (from ten of 32 respondents) concerning the unusable kitchen features. High and low seated showers, multimode bathing fixtures and grab bars were suggested design solutions from six of 32 respondents for the unusable bathroom features. Ten of 32 respondents had problems with storage spaces and suggested a remote-controlled storage system that could have movable and adjustable heights of shelves. There were also other design suggestions by the respondents such as a main entrance without steps, camera installation at entrance doors, a lighted door bell, a push button power door, wide interior doors, motion activated lighting, audible alarms and programmable thermostats.

23.3.2 Institutional Living Environment

Factors Pushing the Institution Respondents to Move

Sixteen institution respondents had been voluntarily relocated and described three factors pushing them to move to alternative accommodation. The first factor, according to 4 of 16 respondents, was the institution's capability of supporting older people's independence, autonomy, and control through the appropriate designs and dimensions of the units to accommodate changing needs. Especially, as researchers stated that the most hazardous room within a living environment is the bathroom (Bakker, 1997), in the institution, the respondents also highlighted their previous bathroom problems. However, the nursing home currently allows them to remain as independent and safe as possible through the use of grab bars and an emergency device in the shower (Figure 23.3). The lower living cost was stated by seven of 16 respondents as the second factor in moving to the institution. The important aspect of this statement is the gender difference. Rather than the females, all of the male respondents dealt with ageing in place from a financial perspective. Five of 16 respondents identified physical support and environmental safety as the third factor. They reported that the institution is designed for comfort (for example, heating, bathroom and power sockets).

(a)

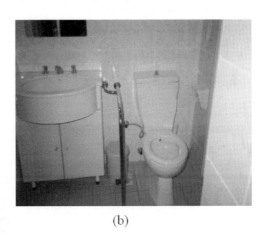
(b)

Figure 23.3. A bathroom example of (a) an emergency device and (b) a grab bar (by the author)

Beyond the environmental factors, gender characteristics also play a significant role in relocation. While research on relocation has primarily focused on characteristics of moving and individual responses to relocation, minimal attention has been given to the process of and desire for moving (Young, 1998). Thus, this study analysed whether the respondents moved voluntarily or compulsorily to this institution. There is a statistically significant relationship between gender and desire to move (x^2=16.000, df=1, α=0,01, two-tailed). All of the 7 male respondents moved compulsorily after their spouse's death, whereas all females moved voluntarily and preferred same-age companionship. The institution provides them with frequent contact with larger social networks. In Turkey, according to the older people, frequency of interactions is closely related with feelings of satisfaction with oneself and life.

Spatial Problems Confronted in the Previous Living Environment
All 16 respondents reported that the architectural design features, services, physical surroundings and leisure facilities in the institution provide them with a better daily living environment than their previous houses. This implies that with increasing age and urbanisation, the Turkish older people would be more receptive to alternative accommodation, if good services, well-designed spaces and high quality

of facilities were provided. Seven of 16 respondents stated that the housework was problematic in their previous houses because of long/narrow hallways, lack of clear spaces and unusable features. The other seven of 16 respondents identified maintenance problems as a factor pushing them to move. The rest faced problems with stairs, as was reported in the previous section by eight of 32 home respondents. Similarly, the spatial problems that the institution respondents had confronted in their previous living environments were closely related to the inappropriateness of their house size.

Desired Universal Design Features

Finally, the respondents were asked about the desired universal design features. The author guided them with the same questions as in the home environments. As all respondents defined the institution as the most appropriate living environment, where necessary support and proper dimensions for use, approach and access are provided, they listed their current living conditions as having the desired design features. They identified four guiding design concepts that offer them the possibility of being safe and independent as long as possible: less physical effort; usability in the kitchen and bathroom; accessibility in storage space without excessive reaching, twisting, and bending; and accommodation of a wide range of abilities. Regarding the concept of less physical effort, five of 16 respondents pointed out the importance of accessible hallways connecting all spaces comfortably and door handles that were operable without twisting. Two of 16 respondents described the usable kitchen and bathroom design as their preference with adaptable cabinets for kitchens and showers instead of bathtubs in bathrooms. Six of 16 respondents highlighted the necessity for accessible storage spaces within easy reach in daily living environments. For three of 16 respondents, the accommodation of a wide range of abilities and the provision of clearances at doors, toilets and turning spaces were the other significant concepts.

23.4 Discussion

All 48 respondents wanted to age in place where they were. The comparison between the home respondents' and institution respondents' attitudes toward ageing in place is essential for further investigations and future developments in home modifications, rehabilitations, financial planning and disability prevention services. Moreover, the verbal responses of the older people and the photographs of their living environments constitute valuable information for designers to incorporate older people needs in the housing design process. Rather than generalising the results for the whole population, this cultural study proposes further elaboration of results by searching for a response to how universal design can be employed. The key issue is to promote universal housing, which means to make living environments habitable, accessible and usable regardless of the disabilities that may occur during the ageing process. Findings are summarised in a key features list (Table 23.2).

Table 23.2. Key features list to increase function and usability

House Area	Universal Housing Feature
Entrance	Main entrance without steps.
	Camera installation at the entrance door
	Lighted door bell
	Push button power door
General Interior	A remote-controlled storage system
	Movable and adjustable heights of shelves
	Wide interior doors
	Operable handles for doors
	Accessible storage spaces within easy reach
	Clear floor areas at doors, toilets and turning spaces
Hallway	Short wide hallways
	An accessible hallway connecting all spaces
Kitchen	An electrically operated counter top
	Pull-out work boards
	Adaptable cabinets
Bathroom	High and low seated shower
	Multimode bathing fixture
	Grab bars
Environmental Controls	Motion activated lighting
	Audible alarms
	Programmable thermostats

This list offers initial design options for those who are experiencing spatial difficulties and are willing to rehabilitate their homes. These features allow the older people to accommodate their needs safely, survive without the need to relocate and make their day-to-day activities easier and home tasks possible. They are also low-cost solutions. However, this 21-item features list in its current form is an initial step. It should be investigated further, which forms the future research agenda of the study. Incorporating as many universal features as possible is essential for designers to satisfy the changing needs of the older people. For this purpose, the Center for Universal Design (2007b) defined gold, silver and bronze universal design features that should be included in a house to achieve a higher level of inclusivity. Designing supportive, adaptable, and accessible daily living environments with innovative features is the main goal of universal housing, and such an environment opens up the possibility of attractive and enjoyable living. Accordingly, a universal house should include the gold key features regarding entrances, interior circulation, bathrooms, kitchens, garage, laundry, storage, hardware, sliding doors and windows (Young and Pace, 2001; Center for Universal Design, 2007a, 2007b).

23.5 Conclusions

Designing for an ageing population and people with diverse abilities has been a prominent part of the universal design movement. As people age, their needs in living environments will also change. "For housing to adequately address these needs all home design must recognise and accept that being human means there is no one-model individual whose characteristics remain static through their lifetime" (Center for Universal Design, 2000). Examining the design literature on universal design showed that there were similar case studies on universal design housing features. Unlike these studies, this study analysed the universal housing issue from the Turkish older people's point of view. It provides directions for future studies on how to identify universal features in daily living environments systematically and how to further evaluate possible universal housing solutions for ageing in place. Moreover, this is ongoing research and the author is currently working on comparing these findings across cultures.

In conclusion, it is important to interpret the findings from two points of view. First is the user-consciousness of ageing in place and second is the designers' awareness of the older people's needs and their attitudes toward alternative living environments. Whether users are young, old or somewhere in-between, they should be concerned with the question of whether or not their houses will respond to their changing needs and disabilities of the ageing process. They should be conscious of any physical limitation that they might experience for their entire life in their own homes. The second important issue is the designers' key role in creating daily living environments. They should be encouraged to engage with the everyday challenges of ageing populations. As in the study, the apartment living pattern observed in Turkey is not capable of responding to the requirements of ageing. The lack of consideration of the requirements of the Turkish older people resulted in the dissatisfaction with their current living environments. In this respect, it would be advisable to take a user-centred approach and provide alternative design solutions suitable for older people's needs both in Turkey and in other countries all over the world.

23.6 Acknowledgments

This research was done within the framework of ID 531 course "Models and Methods of Ergonomics" in the Middle East Technical University, Department of Industrial Design. I would like to thank to the course instructor, Assoc. Prof. Cigdem Erbug, for her broad-minded comments.

23.7 References

Bakker R (1997) Elderdesign: designing and furnishing a home for your later years. Penguin Books, New York, NY, US

Center for Universal Design (2000) Affordable and universal homes: a plan book. North Carolina State University, NC, US

Center for Universal Design (2007a) Residential rehabilitation, remodelling and universal design. North Carolina State University, NC, US. Available at: www.design.ncsu.edu/cud/ (Accessed on 23 September 2007)

Center for Universal Design (2007b) Gold, silver and bronze universal design features in houses. North Carolina State University, NC, US. Available at: www.design.ncsu.edu/cud/ (Accessed on 23 June 2007)

Coleman R, Pullinger DJ (1993) Designing for our future selves. Applied Ergonomics, 24: 3

Demirbilek N, Demirkan H (2004) Universal product design involving elderly users: a participatory design model. Applied Ergonomics, 35: 361–370

Dewsbury G, Clarke K, Rouncefield M, Sommerville I, Taylor B, Edge M (2003) Designing acceptable 'smart' home technology to support people in the home. Technology and Disability, 15: 191–199

Fozard JL, Rietsema J, Bouma H, Graafmans JAM (2000) Gerontechnology: creating enabling environments for the challenges and opportunities of ageing. Educational Gerontology, 26: 331–344

Imamoglu EO, Imamoglu V (1992) Housing and living environments of the Turkish elderly. Journal of Environmental Psychology, 12: 35–43

Imamoglu O, Kılıc N (1999) A social psychological comparison of the Turkish elderly residing at high or low quality institutions. Journal of Environmental Psychology. 19: 231–242

Imrie R, Hall P (2001) Inclusive design: designing and developing accessible environments. Spon Press, London, UK

Iwarsson S, Stahl A (2003) Accessibility, usability and universal design-positioning and definition of concepts describing person-environment relationships. Disability and Rehabilitation, 25: 57–66

Kort YAWS, Midden CJH, Wagenberg AFV (1998) Predictors of the adaptive problem solving of older persons in their homes. Journal of Environmental Psychology, 18: 187–197

Mace RL (1988) Universal design: housing for the lifespan of all people. US Department of Housing and Urban Affairs. Washington D.C., US

Marshall R, Case K, Porter JM, Sims R, Gyi DE (2004) Using HADRIAN for eliciting virtual user feedback in 'design for all'. Proceedings of the Institution of Mechanical Engineers Part B: Engineering Manufacture, 218: 1203–1210

Mueller JL (1997) Case studies on universal design. The Center for Universal Design, Raleigh, NC, US

Ostroff E (1989) A consumer's guide to home adaptation. Available at: www.adaptenv.org (Accessed on 20 June 2004)

Ostroff E (2001) Universal design: the new paradigm. In: Preiser FEW, Ostroff E (eds.) Universal design handbook. McGraw-Hill, New York, NY, US

Pynoos J (1992) Strategies for home modification and repair. Generations, 16: 21–25

Story MF (1998) Maximizing usability: the principles of universal design. Assistive Technology, 10: 4–12

Wagnild G (2001) Growing old at home. Journal of Housing for the Elderly, 14: 71–84

Wilder Research Center (2002) Practical guide to universal home design: convenience, ease and livability. Available at: www.tcaging.org/downloads/homedesign.pdf (Accessed on 5 November 2005)

Young HM (1998) Moving to congregate housing: the last chosen home. Journal of Aging Studies, 12: 149–165

Young LC, Pace RJ (2001) The next-generation universal home. In: Preiser FEW, Ostroff E (eds.) Universal design handbook. McGraw-Hill, New York, NY, US

Chapter 24

Universal Design Patterns and Their Use in Designing Inclusive Environments

H. Froyen

24.1 Introduction

In the epistemological fault zone that designers are currently traversing, the renaissance ideal of *designing for the universal person* is gradually being transformed into a post-modern goal of *universal designing for the variety of real people in real situations*. The notion of the normality of users of person-made environments is becoming less exclusive and gradually including people with permanent, temporary and/or situational functional limitations throughout all phases of the life cycle.

Three major experiences and observations play a fundamental role in our overall research project (Froyen 2004a; 2004b; Froyen and Herssens, 2006):

1. Conflicting interests: The needs and requirements of specific user groups can be analysed and documented, but the crux of the proposed universal design (UD) pattern approach lies in the integration of all these different and often conflicting requirements in sets of coherent UD patterns, which resolve all possible demands.
2. Participation of users/experts: Users with specific mental and/or physical impairments often have unique coping strategies to deal with disabling built environments and objects (Ostroff, 1997). They know better than anyone else which design interventions could best improve their functionality and their comfort. Academics and professionals need to learn from these experiences and insights, and they need to attentively co-operate with these users/experts.
3. The notion of universal designing (Salmen, 2001) as an ongoing process , rather than universal design as a single act should play a central role. In a similar fashion, the European term 'design for all' should actually be interpreted as 'design for more', by stretching the fit each time in every design action to include more users in an optimal way.

The real goal of fundamentally and inclusively designing for everyone requires that researchers and designers return to the broad field of thematic buildings and thematic urban spaces, to the structure of the everyday built environment and to the users. An important prerequisite for this methodical research of the systematic elimination of handicap situations in the (everyday) built environment is that buildings are not viewed as autonomous objects, but as entities in a dynamic social-spatial fabric.

For Christopher Alexander (2003), the entire community of users – both past and present – works by means of innumerable large and small, formal and informal 'structure preserving transformations' of a built environment to provide accommodation in a meaningful and versatile way for human needs and aspirations. He developed a *pattern language* (Alexander *et al.*, 1977) as a generic system to nurture and to guide human building processes. Both Alexander's well known pattern language and the more recent specific initiative of a group of software architects to develop collective *design patterns* (Gamma *et al.*, 1995) provide inspiration for our methodological approach to universal design. Successful collective web-based projects in open content (OC) further inspire us in the development of models of UD patterns and in the search for strategies for implementation.

In addition to the UD patterns, briefly presented in this chapter, the overall research project comprises six complementary components of the methodological approach. Three of these components (empirical research, simulations, and users/experts in collective design processes) all contribute to the expansion of the universal design knowledgebase and to the development of the UD patterns. Two further components (integral quality control and post occupancy evaluation (POE)) also contribute to the continual improvement of universal accessibility and usability. Feedback from facility management and from evaluation processes is channelled back into the UD patterns. The sixth and final component (universal design education and research) is viewed as a primary construction block for the long-term development of a rich academic and professional culture and tradition in support of the new universal design paradigm. In this paper, only the elaborated model construct of UD patterns will be further explored.

24.2 Users

In the past decades the conventional requirement for physical accessibility was primarily based on wheelchair accessibility and the needs of people with visual limitations. The design parameters for these two user groups have been well documented. For the great majority of the remaining physiological, motor, sensory, neurological, anthropometrical, mental and psychological functional limitations, the relevant design data must be distilled from the results of empirical medical, paramedical and architectonic research.

The International Classification of Functioning, Disability and Health of the World Health Organisation (ICF, 2001) is an important potential source of empirical data in connection with environment-related human activities and participation. It documents human functioning and human disabilities under four large categories: body functions, body structures, activities and participation, and environmental factors. World-wide data in 889 categories (d110 through d999) relating to human functioning, disability and health are collected and processed in a standardised way. The fundamental structuralistic ICF classification of the WHO could in the long term form a solid basis for registering, describing and structuring problem definitions (conflicts) in the UD patterns. However, the classification is currently based on medical conditions and it lacks focus on the complex two-way dynamic interaction between users and the built environments.

An important goal in our UD pattern research project is to find ways of dealing with the functional effects of disabilities without getting tangled in medical jargon or non-relevant and often compromising confidential medical information. Therefore, in the problem definition (conflicts) section of our proposed UD patterns, we elaborate on an adapted scheme of relevant human functional characteristics that are important and need to be considered when designing for human use. Functional characteristics listed in 'the enabler' (Steinfeld et al., 1979), illustrated in Figure 24.1, are integrated here in the five-chapter scheme represented in the Dutch publication Geboden Toegang [Access advisable] (Stichting Nederlandse Gehandicaptenraad, 1986).

It is important to emphasise that the compiled list of human functional characteristics (deductive) is only intended for the construction of a working UD pattern model. In the further operationalisation of UD patterns, we aim for the detection of relevant misfits (conflicts) in concrete interactions between users and built environments (handicap situations) through post occupancy evaluation of the existing built environment, and by means of closely related architecture-specific empirical research (inductive).

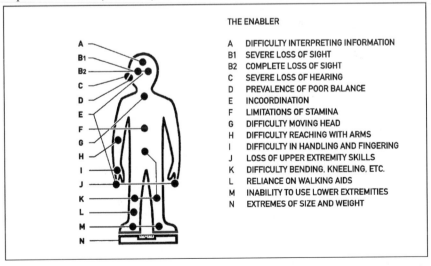

THE ENABLER

A   DIFFICULTY INTERPRETING INFORMATION
B1 SEVERE LOSS OF SIGHT
B2 COMPLETE LOSS OF SIGHT
C   SEVERE LOSS OF HEARING
D   PREVALENCE OF POOR BALANCE
E   INCOORDINATION
F   LIMITATIONS OF STAMINA
G   DIFFICULTY MOVING HEAD
H   DIFFICULTY REACHING WITH ARMS
I   DIFFICULTY IN HANDLING AND FINGERING
J   LOSS OF UPPER EXTREMITY SKILLS
K   DIFFICULTY BENDING, KNEELING, ETC.
L   RELIANCE ON WALKING AIDS
M   INABILITY TO USE LOWER EXTREMITIES
N   EXTREMES OF SIZE AND WEIGHT

Figure 24.1. The enabler (Steinfeld et al., 1979)

24.3 Built Environments

In preparation for the following model construct of a UD pattern we progress step by step through four phases of decomposition to the final selection and delineation of the smallest possible relevant elements (toilet, door handle or staircase) or aspects (daylight requirements, perceptible information or way-finding). The social-spatial element or aspect then constitutes an object for working out and formulating patterns (conventional patterns for modal users, average, standard), or in this case of UD patterns specific conflicts / resolution data to facilitate the systematic elimination of handicap situations for everyone in built environments.

Phase 1: Selection and delineation of a social-spatial entity from the global built environment.

Global built environment ➜ *Selected social-spatial ENTITY*

The movement and the activities of people in the stages of a travel chain occur in public spaces and in public infrastructure which is part of the global built environment. Empirical research could indicate, for example, that in a certain train station the access of a significant number of users is hindered. We can then scientifically isolate that train station from the global built environment as a definable and analysable social-spatial entity.

Phase 2: Decomposition of the selected entity into autonomous social-spatial settings (arrival outside, the main concourse inside, platforms).

Selected social-spatial entity (for example, a train station) ➜ *relatively autonomous social-spatial settings, quite clearly delineated (arrival outside, the main concourse inside, platforms)*

Figure 24.2 gives a graphic image and synthesis of five distinct social-spatial settings inside the selected entity of a train station and its environment. Here these empirically documented social-spatial settings are specifically designated as stages of a travel chain (Finnish Government, 2003).

Phase 3: Further decomposition of the selected settings in clusters of patterns (vertical traffic inside the train station, such as one cluster of fixed stairs, steps, moving walkways, ramps, escalators, platform lifts, lifts).

Settings ➜ *cluster of UD patters (Vertical traffic, for example, the entirety of fixed stairs, moving walkways, ramps, escalators, platform lifts, lifts)*

Each of the five social-spatial settings can now be unfolded further to the level of the smallest possible social-spatial cluster of UD patterns.

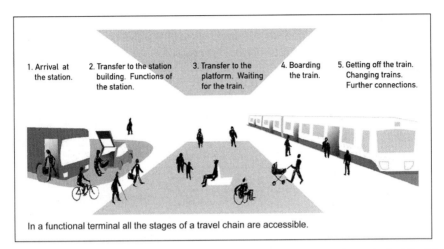

1. Arrival at the station.

2. Transfer to the station building. Functions of the station.

3. Transfer to the platform. Waiting for the train.

4. Boarding the train.

5. Getting off the train. Changing trains. Further connections.

In a functional terminal all the stages of a travel chain are accessible.

Figure 24.2. Selected social-spatial ENTITY (in this example, a train station). (Finnish Government, 2003. Illustration by J. Kunnas).

Phase 4: In conclusion, within the formulated cluster of UD patterns, one or more of the autonomous UD patterns ('atoms of the environmental structure') is isolated for further research.

Cluster of UD patters ➔ autonomous single UD patters ('atoms of the environmental structure', for example, a staircase)

If we now, for example, focus on the smallest possible autonomous social-spatial setting within that large entity of a train station, then the characteristics of the element 'staircases in public buildings' can be temporarily removed from the complex social-spatial context and investigated with a view to a problem definition (conflicts) and to an optimal morphological/technological design solution (resolution).

24.4 Universal Design Patterns

UD patterns form the core elements of the proposed methodological approach. They provide relevant information in a structured way both about problems (conflicts) that are experienced in handicap situations by users (with specific permanent limitations or non-permanent temporary or situational limitations) and empirically supported architectural/technological solutions (resolutions). Since the documented solutions are always tentative and a large amount of room remains open for innovative concepts, we prefer to describe the UD pattern approach as an adapted problem definition method. By explicitly giving attention in the UD patterns to the problem definitions (conflicts), we differ fundamentally from the many prescriptive legal ordinances and building regulations that are presently used in design and building processes.

UD patterns accurately capture descriptions of the 'why' for each design parameter. By their descriptive nature, they also contribute to a broad user-oriented design and building culture that complements the prescriptive national and European laws and norms. For the development and the continual updating and improvement of UD patterns, we propose that conventional empirical research be combined with peer review (users/experts) and with a broad exchange of OC information and communication via the Internet.

In the complex interface zone between users and built environments, problems in terms of a number of disabling handicap situations (conflicts) can be detected and can be analysed in specific UD patterns as shown in Figure 24.3.

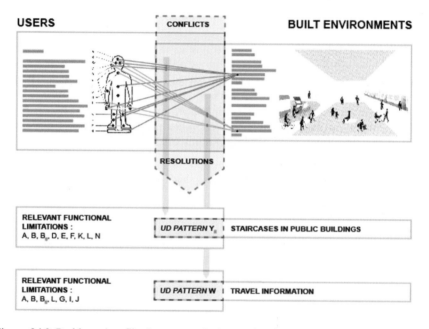

Figure 24.3. Problems (conflicts) can occur in the overlapping interface zone between users and built environments

24.5 Universal Design Pattern Model

The majority of the distinct UD Patterns are hierarchically associated and interrelated with larger-scale and/or smaller-scale related patterns. For example:

LARGEST SCALE UD PATTERNS
UD Pattern X: Vertical traffic of users in public buildings. General.

INTERMEDIATE SCALE UD PATTERNS
UD Pattern Y_1: Risers in public buildings

UD Pattern Y_2: Staircases in public buildings
UD Pattern Y_3: Fixed ramps in public buildings
UD Pattern Y_4: Horizontal and sloped moving walkways in public buildings
UD Pattern Y_5: Escalators in public buildings
UD Pattern Y_6: Platform lifts in public buildings
UD Pattern Y_7: Lifts in public buildings.

SMALLEST SCALE UD PATTERNS

UD Pattern Z_1: Geometry and material of risers and treads
UD Pattern Z_2: Dimensions of the risers and treads of steps and stairs in public buildings
UD Pattern Z_3: Handrails for steps and stairs in public buildings
UD Pattern Z_4: Step markings for steps and stairs in public buildings
UD Pattern Z_5: *etc.*

The UNIVERSAL DESIGN PATTERN model elaborated below in Figure 24.4 presents a simplified static model of a possible interactive digital version.

UD Pattern Y_2: Staircases in public buildings

• **Introductory paragraph with references to related UD Patterns to which this specific UD Pattern serves as a supplement.**

Vertical traffic of users in public buildings is made possible and is facilitated by stairs, fixed ramps, horizontal or sloped moving walkways, escalators, platform lifts and lifts. This cluster of architectural and/or technological elements, and their mutual relation and position, will be thoroughly analysed and documented in one overarching **UD Pattern X: Vertical traffic of users in public buildings. General.**
The majority of visitors and users can make optimal use of the conventional risers described in **UD Pattern Y_1: Risers in public buildings**, and of fixed stairs, described further in **UD Pattern Y_2: Staircases in public buildings.** Certain persons with permanent, temporary or situational limitations can only make use of fixed ramps: **UD PATTERN: UD Pattern Y_3: Fixed ramps in public buildings**, horizontal and sloped moving walkways **UD Pattern Y_4: Horizontal and sloped moving walkways in public buildings**, escalators: **UD Pattern Y_5: Escalators in public buildings**, platform lifts: **UD Pattern Y_6: Platform lifts in public buildings**, or lifts, as analysed and described in **UD Pattern Y_7: Lifts in public buildings.**

Figure 24.4. (a) Problem definitions and the architectural/technologica solutions

---

• Problem Definition (CONFLICTS)

Risers and stairs for vertical traffic are certainly usable by many, yet they hinder all users to a greater or lesser degree because of the extra physical effort required when going up, and because of the danger of tripping, losing balance or falling. 'Stair climbing demands a rate of energy expenditure that is higher than any other routine daily activity and is comparable to heavy physical labour' (Templer, 1992).

The usability of stairs for visitors and users with permanent, temporary or situational limitations is partially dependent on their intrinsic disabilities but also partially on extrinsic environmental factors relating to the way in which the steps are designed, constructed and maintained.

In the following paragraphs of this specific model UD Pattern 'Y$_2$: Staircases in public buildings' there will be a systematic investigation and definition of the different facets of problem formulation (CONFLICTS) for the vertical traffic of visitors and users with various permanent, temporary or situational physical/mental capacities and limitations (Category 0.0 through category 5.0).

- 0.0 Modal users (average, standard). This includes users who are tired, pregnant, stressed, ill or injured, undergoing medical treatment, under the influence of alcohol or drugs, as well as travellers with a pram, with baggage or with heavy or sizeable objects.

- 1.0 Users with neuromusculoskeletal and movement related functional limitations.

Functional limitations and handicap situations can result from problems in the area of movements, actions and mobility.

- 1.01 Lack of co-ordination

- 1.02 Impairment of the walking function. Reliance on walking aids. Incorrectly designed stairs prevent certain categories of users with a limited walking function from using them. Risers that are too high form a barrier and open risers can cause a (dragging) foot to get caught. High risers and shallow treads also make use more difficult and increase the danger of falling. Absence or inadequacy of handrails hinders people with an impaired walking function because of the lack of needed support.

- 1.03 Lack of the walking function. Inability to use lower extremities. Reliance on wheelchair and supports. Cannot make use of stairs independently. These people must occasionally and in emergency situations be carried up or down the stairs by one or more people.

- 1.04 Impairment of the arm or hand function. Difficulty reaching with arms. Difficulty in handling and fingering (arthritis, arthrosis). Loss of upper extremity skills. Possible use of the handrail is determined by the height and ergonomic form of the handrail and by its presence to the left or right of the user.

- 1.05 Difficulty moving head

- 1.06 Difficulty bending, kneeling, *etc.*

- 1.07 Prevalence of poor balance. Absence of handrails, non-ergonomically constructed handrails, handrails that are only present above the steps, and do not continue on horizontally in the area before and after the stairs, prevent users with a loss of balance from using stairs in a safe and efficient manner.

Figure 24.4. (b) Problem definitions and the architectural/technologica solutions

- **2.0 Users with sensorial limitations.** Functional limitations and handicap situations can result from problems in the area of perception.
- 2.01 Auditive limitations: decreased hearing function. Severe loss of hearing
- 2.02 Lack of hearing function: deaf
- 2.03 Visual impairment: Severe loss of sight. Insufficiently lit and poorly marked stairs, stairs with trash, steps without a colour contrast, lack of or inefficient tactile warning strips, lack of handrails as natural guidelines, prevent users with visual limitations from using stairs in a safe and efficient manner.
- 2.04 Lack of visual perception: blind. Poorly located and poorly marked stairs, lack of or inefficient tactile warning strips, lack of handrails as natural guidelines, handrails that are only present above the steps and do not continue on horizontally in the area before and after the stairs, prevent users without vision from using stairs in a safe and efficient manner. People with visual limitations cannot detect elevated obstacles with a walking cane. Non-delimited floor areas under open stairs cannot be detected by visually impaired people using a walking cane, which can lead to collisions with stairs and to accidents.
- **3.0 Users with organic defects.** Functional limitations and handicap situations can result from problems in the area of required physical effort.
- 3.01 Impairment of the heart function. A lack of platforms, wrongly-dimensioned risers, a lack of handrails, or non-ergonomically constructed handrails, place an unnecessary strain on the limited stamina of persons with physical limitations and/or organ defects.
- 3.02 Impairment of the lung function and of the bronchial tubes. Stairwells that are poorly ventilated and/or dusty are a hindrance to users with an impairment of the lung function and of the bronchial tubes.
- 3.03 Impairment of the kidney function
- 3.04 Limitations of vigour and stamina (energy level)
- **4.0 Users of exceptional size.** Functional limitations and handicap situations can result from problems in the area of anthropometrics.
- 4.01 Small stature. High risers and high handrails make taking the stairs more difficult and more dangerous for people with a small stature.
- 4.02 Large stature. Too little headroom and handrails that are too low can be a hindrance to users with a large stature when taking the stairs.
- 4.03 Extremes of size and weight. Narrow stairs and narrow winding stairs, stairs with an unfavourable riser/tread ratio, a lack of handrails or non-ergonomically constructed handrails, a lack of platforms or platforms that are too small, prevent users of exceptional size and weight from using stairs efficiently and safely.

Figure 24.4. (c) Problem definitions and the architectural/technologica solutions

- 5.0 Users with mental and/or psychological limitations. Functional limitations and handicap situations can result from problems in the area of cognitive, neurological and psychological capacities.

- 5.01 Mental limitations. Difficulty interpreting information. Mental limitations and non-congenital brain damage (acquired brain injury) can lead to a lack of skills and to difficulties in carrying out simple tasks such as taking the stairs. Many people with mental limitations and acquired brain damage have little or no sense of orientation, even inside buildings (Dumortier, 2002). Many have a limited sense of balance and have problems with estimating height and depth. Lack of handrails or non-ergonomic handrails, handrails that are too short and do not continue on horizontally in the area before and after the stairs, can make it more difficult or impossible to take the stairs (Dumortier, 2002).

- 5.02 Psychological limitations. Psychological limitations including confusion, stress and phobias (acrophobia) can be made worse by free-standing stairs without guiding enclosing walls. Complicated way-finding to the stairs and poor recognisability can lead to increased confusion. Bad lighting and poor maintenance increase the danger of falling and of fear.

• Results of empirical research

(In this section of the UD Pattern, a bridge of explicit and verifiable logical reasoning will be built up. The gradual crafting of an Architectural / Technological solution (RESOLUTION) is supported by empirical data, relevant remarks of users/experts, prototypes and case-studies, *etc*.)
REFERENCES, specific for UD Pattern Y2: Fixed stairs in public buildings. A full list of references will be added, but due to space constraints in this paper they were not included here.

• Architectural / Technological solution (RESOLUTION)

(In this section of the UD Pattern a best possible Architectural / Technological solution (RESOLUTION) will be presented and detailed. The graphic / word description of relevant details of the staircase can be marked with the corresponding codes referring to the appropriate paragraph in the Problem Definition (CONFLICTS) section above, in this case 1.02, 1.04, 1.07, 2.03, 2.04, 3.01, 4.03, 5.01 and 5.02)

• Closing paragraph with references to related UD Patterns that supplement and round out this specific pattern

The following UD Patterns provide detailed information and round out the above UD Pattern Y$_2$: Staircases in public buildings: **UD Pattern Z$_1$: Geometry and material of risers and treads** and **UD Pattern Z$_2$: Dimensions for the risers and treads of steps and stairs in public buildings.** Also handrails **UD Pattern Z$_3$: Handrails for steps and stairs in public buildings.** Visual and tactile markings are detailed further in: **UD Pattern Z$_4$: Step markings for steps and stairs in public buildings.**

Figure 24.4. (d) Problem definitions and the architectural/technologica solutions

The above approach for Problem Definitions (CONFLICTS) and the Architectural/Technological solutions (RESOLUTION) is purely tentative and illustrative. The data are exclusively based on publications at hand. When working out the details and making this model construct for a UD Pattern operational, broader multidisciplinary research and peer review by users/experts will be of essential importance. In an open web-based forum, scientific research results will be continually tested by the experiences and findings of users/experts and of professionals.

24.6 Conclusions

For the last five years (2002-2007) the described methodological approach has systematically been tested and improved in the design studio, with a yearly average of 55 third year students at PHL Department of Architecture, Hasselt, Belgium (Froyen, 2004a). The work in progress yields satisfying results in the academic field, but there is no evidence that the UD design pattern approach will also be successful on a larger scale in the social and in the professional realm. Critical analysis of similar running projects reveals some of the obstacles that have to be overcome.

The American project 'The Anthropometrics of Disability' and the Dutch Delft University's 'Antropometrisch Informatie Systeem' (AIS) are two advanced and inspirational international initiatives, with goals similar to our methodological approach. Most closely related to our presented UD pattern, however, is a more modest local initiative, developed in 2000 by D-science lab, an Antwerp Design Sciences Research Center in Belgium. Their AGEtree database is still online, but unfortunately no longer updated, due to a lack of financial and logistic support. This nearby example highlights in a dramatic fashion the biggest obstacle our elaborated methodological universal design tool will probably also be faced with, namely the substantial and long-term off-line academic, professional and material support needed to develop and to permanently update and improve an on-line database and a related web-based discussion forum.

Moreover, well-engineered UD patterns can substantially contribute to, but not guarantee, universal access. As stated in the introduction, six complementary components of our overall methodological approach work together with UD patterns as generative techniques in the dynamic processes of the creating and adaptating of enabling environments.

Neither UD patterns nor the UD principles are intended as universal criteria for good architecture. Conventional aspects such as aesthetic qualities, cultural-historical values, sustainability, safety and economic values must also be included in every design (Froyen, 2003). The universal design paradigm neither hinders nor promotes any specific architectural style. Instead, it puts an emphasis on a user-centred approach rather than a design-centred attitude (Whitney, 2003). It might be useful to start a broad social/academic discussion about sustainable, elegant, and enabling built environments as opposed to conventional aesthetic buildings.

24.7 References

Alexander C, Ishikawa S, Silverstein M, Jacobson M, Fiksdahl-King I, Angel SA (1977) A pattern language. Oxford University Press, New York, NY, US

Alexander C (2003) New concepts in complexity theory, arising from studies in the field of architecture. Available at: www.katarxis3.com/SCIENTIFIC%20INTRODUCTION.pdf (Accessed on 15 November 2004)

Dumortier D (2002) Van een ander planeet. Autisme van binnen uit. [From another planet. Autism from inside out]. Houtekiet, Antwerp/Amsterdam, The Netherlands

Finnish Government (2003) Towards accessible transport, accessibility strategy of the Ministry of Transport and Communications. Ministry of Transport and Communications, Helsinki, Finland

Froyen H (2003) Universal design education. In: Dujardin M, Dua I (eds.) Universal design education. In: Proceedings of the UD Education Contact Forum, Brussels, Belgium

Froyen H (2004a) Universal design education. In: Spiridonidis C (ed.) Monitoring architectural education in European schools of architecture. EAAE Transactions in Architectural Education, 19: 125–132

Froyen H (2004b) Universal design, or systematic elimination of handicap situations in the built environment. In: Proceedings of the 5th FTW PhD Symposium. University of Ghent, Belgium

Froyen H, Herssens J (2006) Belgian universal design education project (2001-2006). In: Proceedings of the 6th International Conference on Universal Design of Buildings: Tools and Policy. Brussels, Belgium

Gamma E, Helm R, Johnson R, Vlissides J (1995) Design patterns, elements of reusable object-oriented software. Addison-Wesley, Reading, MA, US

ICF (2001) International classification of functioning, disability and health. World Health Organization, Geneva, Switzerland

Ostroff E (1997) Mining our natural resources: the user as expert. IDSA, 16(1): 33

Salmen PS (2001) U.S. accessibility codes and standards: challenges for universal design. In: Preiser FEW, Ostroff E (eds.) Universal design handbook. McGraw-Hill, New York, NY, US

Steinfeld E, Schroeder S, Duncan J, Faste R, Chollet D, Bishop M (1979) Access to the built environment. A review of the literature. Government Printing Office, Washington D.C., US

Stichting Nederlandse Gehandicaptenraad (1986) Geboden Toegang. [Access advisable]. Utrecht, Belgium

Templer J (1992) The staircase. Studies of hazards, falls, and safer design. MIT Press, Boston, MA, US

Whitney P (2003) Design in a global world. Interview with Professor Patrick Whitney on 4 November 2002. EAAE News Sheet, 66(June): 17–25

Chapter 25

User Friendly Living Environmental Research and Design for Older People

J.Y.C. Kwok and K.C.H. Ng

25.1 Introduction

It is obvious that we are entering into an ageing society. Taking Hong Kong as an example, in the year 2030, it is projected that 22% of the total population will have reached 60 years old or above. In Hong Kong, there is still a lack of inclusive and comprehensive planning for society to meet this change in demography. Nowadays, the percentage of educated older people is becoming significant. It is also projected that 20 years on 63.6% of the elderly population will have received secondary education level or above. Obviously, people who are better educated will have different expectations in life as compared with the present cohort of older people. They may keep lifestyles with variety even when they become older.

Recent gerontological geography researches focus on discussions of living environment design and planning in a multidisciplinary context. In *Placing Ageing, Positioning in the Study of Older people*, Kearns and Andrew (2005) state clearly that discussions on city planning today should enter into the era of "post-medical geography of health" *i.e.* it is not sufficient to plan an environment where only responsive type of medical services are provided. The sustainability of the physical and psychological health of the citizens is a broad concern which involves serious consideration in a multidisciplinary context.

The definition of health offered by World Health Organisation was "a state of complete physical, mental and social wellbeing" (Bond and Corner, 2004). According to Bond and Corner, for the older people, the important components of a good life quality are care from the family (children), social contacts, health, mobility/ability, material circumstances, activities, happiness, youthfulness and living environment.

Following these criteria, we understand living environment planning for older people does not only mean 'housing' planning; a proper living environment should allow people to develop and sustain their personal identity, to initiate human interactions, and to form a community.

Phenomenologist Norberg-Schulz (1985) explains that our actions in the world configure and create spatial orders. In a concrete space people continue to negotiate with ethical, social, economic and aesthetic orders to organize their life. They also search for intimate relationships, to acquire and defend individual/ communal culture and values. The life experience 'gathers' into memories and it will emanate in the place where people live. As one gets old, one's life experience sediments into spatial expressions. De Certeau (1984) is indeed poetic and clairvoyant to say, 'haunted places are the only ones people can live in'. The environment should therefore be able to embody the past, and give people freedom of action in order that the self identity can be sustained.

When the older people are in good health, they will naturally lead an active life and travel widely. Moreover, since the birth rate continues to drop, in the near future, people aged 60 or above may have to continue to work (and to be tax payers). How will our city be prepared for these new social phenomena of active ageing and elderly working population? When a significant number of citizens are older people, the city should re-adjust and re-design its facilities so that this group of senior but active citizens can continue to participate in different aspects of city life.

When one gets old, one will also become more attached to the familiar home environment physically and psychologically. According to Norberg-Schulz (1985), one will withdraw from the 'public world' to the home to recover one's identity. The home is a place where people gather memories of the world. The related institutions and professionals should be able to design a warm and cozy 'interior' in which people can 'gather' their life.

Baltes and Carstensen (1996) propose that the concept 'successful ageing' should be applied to define the term 'ageing'. The two scholars introduce a new model to describe the process of ageing: 1) selection; 2) compensation; 3) optimization. They think that, when people get old, they will reconsider their prioritization of the goals to be achieved in life. In this case, 'selection' means 'the readjustment of individual goals'. When we discuss the 'selection' issue in respect to environment, we may consider that the older people will stay more at home, and/or incline to move about in the immediate neighbourhood. In this respect, the major task of housing providers and city planners should be the designing of a comfortable home which allows the older people to sustain close and frequent contact with relatives and friends, as well as a supportive neighbourhood in which they can lead an active life.

Following the view of Bartes and Carstensen, we see that 'compensations' would be means to compensate for functional deficiency. With 'compensation' people can thus maintain or optimize the capacity to lead a normal and active life. If the health of older people is deteriorating, sufficient auxiliary facilities and services must be provided to enable them to move around.

Baltes and Carstensen state that 'optimisation' refers 'to the enrichment and augmentation of reserves or resources and, thus, enhancement of functioning and adaptive fitness in selected domains'. It is therefore obvious that an 'optimal' living environment is one endowed 'with a stimulating and enhancing quality'. Bartes and Carstensen (1996) further define 'optimising environments' as places where older people are encouraged to conduct physical exercise and motivated to 'increase memory'.

25.2 Research on Older People Oriented Living Environment 2002 - 05

Our research team collated theoretical standpoints from comparative studies on human geography, phenomenological architectural theory, sociology and gerontology as mentioned above. With such perspectives, our research questions are as follows: (1) How do the older people organize their life? What are their expectations and goals? (2) In a city where the problem of ageing is becoming more and more serious, how are we going to prepare and plan the environment to cope with such social change?

25.2.1 Research Process and Findings

The research process was organised in four stages: (1) Questionnaire survey – evaluation of the older people's (aged 60 or above) capacity for activities in different environments; (2) visual simulation modelling workshop for designing ideal interior space; (3) visual simulation modelling workshop for designing ideal neighbourhood space.

Questionnaire Survey

Based on the Lawton (Lawton and Brody, 1969) instrumental activities of daily living scale, we designed a questionnaire survey (2003-04) to gather information about the following aspects: (1) the social background of the informants; (2) the health condition of the informants; (3) the difficulties the informants encountered when they undertook different daily life activities in their interior space as well as in the exterior public space.

Reply samples were collected from 204 respondents who were all 60 years old or above. 79 samples were from male informants while 125 were from female informants. Most of them were retired and lived alone. From the collected data, we understood that most informants were able to read. Many of them claimed that home cleaning was a difficult job. 15% of the respondents reported that their home allowed no space to do simple physical exercise. Concerning the exterior space, 90% of the older people found that they had difficulty in finding their way and moving about: *e.g.* in train stations and when using escalators. Moreover, most of them agreed that recreational space should be built in the immediate neighbourhood, and should be clean and provided with toilet facilities.

Visual Simulation Modelling Workshop

According to Wates (2000), 'models are one of the most effective tools for getting people involved in planning and design. They are particularly useful for generating interest, presenting ideas and helping people think in three dimensions.' Our workshops, using visual simulation models as the principal tool, enabled the users to express in concrete visual and figurative details their suggestions for an appropriate living environment.

Visual Simulation Modelling Workshop for Ideal Interior Space

To start the workshop, the research facilitators introduced to the participants the purposes of the project and explained to them the idea of participatory research and design.

The modelling kit we designed included partition walls, floor board (representing an area of 800 sq.ft.), basic furniture, basic electrical appliances and plants modelled at the scale of 1:10. Each participant (or couple living together) would design (made a model of) his/her/their ideal home interior. The processes included: (1) participants discussing the furniture arrangements; (2) initializing design process with the base board; (3) setting up partition walls with different types of window size; (4) installing furniture;(5) decorating the home with paint, green plants and other gadgets; (6) finalizing the interior design model.

After designing the ideal home, each participant made a presentation. The participants presented the directions and aesthetic preferences of the design, and the aspirations it revealed. The participants thus expressed the ideas of a meaningful life at home.

Facilitator briefing the participants about the purposes and key issues of the workshop

Facilitator discussion with a participant on his everyday living pattern

Participants constructing the Ideal home space in individual or in couple

A participant with his design

A finished model

Figure 25.1. Ideal interior space workshop in progress

Summary of the workshop results included:

- Ideal living unit size: Most participants considered that a unit with size from 400 sq. ft.^{2} to 600 sq.ft.^{2} would be sufficient for a couple to live in.
- Living space enabling individual cultural activities: Participants expressed their wishes for comfortable reading environment with good lighting. Some participants wanted a big table surface for practicing Chinese calligraphy.
- Living with spouse: A participant stated he preferred a twin bed room because sleeping on a single bed he could avoid disturbing his wife when getting out of bed and going to the toilet.
- A place for their Deity: Participants expressed the need to install a miniature altar/shrine to worship the gods and ancestors.
- Accommodating the children and relatives who come home to visit
- Many participants preferred a spare room in the living unit. This room was for accommodating their children, grand children and relatives who visited them. Such aspirations expressed in fact a deeply-rooted Chinese traditional conception of family: *i.e.* "different generations living under one roof".
- Space to enable living with nature: All participants wanted a living unit with large windows and good lighting. They believed a balcony was a space for leisure such as gardening and relaxation such as watching the moon.
- Hygienic facilities: All participants preferred shower to bath. They suggested that the bathroom should be spacious because people could have easy access to it in case they had an accident.

Visual Simulation Modelling Workshop for Ideal Neighbourhood Space

The Research Group has designed a 3-dimensional site planning kit to the scale of 1:500. The site planning kit consisted of components representing basic social institutions, infrastructure and recreational facilities. On a plastic mat a circle of 20 minutes of walking distance (divided into 5, 10, 15 20 minutes concentric circles) was graphically represented. Altogether, we conducted 7 workshops and each lasted for three hours.

The neighbourhood space was by definition a space for the collective community. The facilitators therefore recommended the participants to form groups to take part in this design workshop.

The workshop is divided into two sections: in the first section, the participants, in groups, were asked to use the facilities-amenities checklist to prioritize the facilities and public amenities with respect to proximity to the home block.

The Research Group then set up the plastic mat representing a circle circumscribing the temporal-spatial measure of 20 minutes walking distance. 2 to 6 participants, in a group, started the 'game' by placing the home block at the centre. The group started to build model of a local neighbourhood. They placed the public facilities and amenities according to the preferred distance to the home block (measured in walking duration). They co-operated with each other in making a model of their ideal neighbourhood. When the neighbourhood model was constructed, the participants belonging to one group were invited to present their ideas. Through their explanation, we came to understand the problems of current housing planning and its relation to the inhabitants' daily life as well as their needs, concerns and aspirations.

This was an effective game to get people to discuss and exchange ideas. In the workshop, the group members started to talk and express their preferences, and to discuss with others in order to achieve consensus.

Participants discussing on the relationship between their individual life and the community life with the facilitators

Participants selecting preferred components for constructing an ideal neighbourhood

Facilitators encouraging the participants to design their ideal neighbourhood space

A group of participants with their design

A finished model

Figure 25.2. Ideal neighbourhood workshop participants in action

The workshop results are summarized as follows:

- Proximity of the home block: Six out of the seven workshop groups believed that the adjacent environment around the home block should be tranquil. Lawn with seating facilities should be located around the home block. Shops providing every day necessities were also placed nearby. They preferred covered walkways for connecting their homes to the places mentioned.

- Community centre and medical clinics: Most participants said that a community centre should be found within 5 minutes walking distance from home. Concerning the need of medical services, participants considered different levels of urgency: 1. Chinese herbal medicine shops and western drug stores were placed nearest to the home block; 2. a western medical clinic was a little farther away; 3. the farthest away was a hospital. This spatial order indicated that participants were inclined to look firstly for the cheapest way to cure illness if it was minor.
- Shopping and leisure space: Older people liked a traditional street market. Their perception was that a street market (with stores and hawker booths) would offer richer choice of products while the price would be cheaper than supermarkets and shopping mall type shops. A market street with a small park in the proximity was regarded as a convenient gathering place to meet friends and neighbours.
- Transportation facilities: For circulation paths the older people preferred tunnels or covered walkways on ground level to flyovers. Covered walkways on ground level were their favourite.
- Green environment: All respondents loved a green neighbourhood space.
- Cultural and recreational facilities: Park, community centre, community hall and library were the preferences of the participants. The sportive participants suggested that a larger park should be found in the neighbourhood for morning exercises. A park was suggested to be an ideal place for relatively passive older people to be involved in community life: they could just sit there and watch people go by. The participants thought that universal design should be implemented when planning the cultural and recreational space and facilities.
- Community Life: Most participants intentionally planned the elderly home, elderly centre, library, schools, kindergarten, community hall, children's and youth centres in one cluster. This showed that the participants wished to live happily among other people in a community.

25.3 Conclusions

After completing the different stages of research and analysis, we have drafted in our report a schema explaining the guidelines for designing a good living environment for older people. The schema consists of 60 concepts. They formed together a holistic and continuous living environment from the bed room to the immediate neighbourhood space. The 60 concepts with abbreviated explanations are listed in Table 25.1. (a-e).

Table 25.1. (a-b) Guidelines for designing a good living environment

(A) Entrance area	
Concept	**Guidelines**
Main entrance and lobby	• Easily recognizable design features • Resting spaces for residents for hanging around and meeting up
Vertical linkage	• Good ventilation • Intermediate resting place
Corridors	• Resting places at regular intervals • Wider passage with handrail. It should allow passage of wheelchairs
(B) The Interior	
Concept	**Guidelines**
Holistic home environment	Design of living environment adopts 3 levels of optimization: 1. Core level: basic living space and facilities design 2. Selective level: more detailed living space and facilities designed for different lifestyles 3. Optimum level: living space and facilities expressing good quality of life and freedom of choice
Entrance to individual unit	• Residents have the freedom to decorate the door surface and resting space • Residents individualize their doors to create a sense of belonging and identity • Well lit door front with small window in it to view visitors
Recognition of individual's unit	• Legible signage system is essential
Seats	• Different types of seats for different areas arranged in a well planned manner
Common area in interior environment	• Easily accessible common area

Table 25.1. (c) Guidelines for designing a good living environment

(C) Home Unit	
Concept	**Guidelines**
Living/dining area	• Flexible space for different kinds of family events and social gatherings
Kitchen	• The cooking culture of the elderly should be carefully considered: *e.g.* Chinese love double-stewed soup • Safety measures should be taken
Toilet and bathroom	• Bathroom door can swing both inwards and outwards • Shower has sufficient space for two adults to move about freely
Wet floor shower system	• Non-slippery floor • Preventive device for water splashing out from shower cubicle
Shower with seat	• Allow elderly to enjoy the shower longer
Lever tap and toilet flushing	• For easy operation
Ventilation and heating in bathroom	• To cope with seasonal temperature change
Sleeping area	• Older people do not prefer studio interior • Partition to separate the sleeping area from other areas • Adequate space for twin beds for couples
Bed	• Should be easy to get down and high enough for sitting on
Bedding	• Changes of bedding material according to season
Bedside lamp and low level lighting	• Table or small light next to the bed • Route from bed to toilet should be lit by low level light
Location of bed	• Bed can be next to window but sunlight should not fall directly onto it
Storage	• Storage space should be provided for out-of-season clothes • Furniture for displaying memorabilia
Wall decoration	• Large pin boards and picture rails for displaying photos
Open shelves and surfaces	• Easily accessible open shelves for storage • Low cabinets and deeper windowsill allowing things to be put on surface top
Cool environment	• Electrical fans can be used for cooling instead of air conditioners which may cause rheumatism

Table 25.1. (c) Guidelines for designing a good living environment

(C) Home Unit	
Concept	**Guidelines**
Window	• Windows, besides providing light and ventilation, give passive access to the outside world
Location of window	• Chinese older people prefer south facing windows
Ventilation	• Exhaust fans should be installed in kitchen and toilet
'Room with a view'	• Viewing is a 'subtle and passive means of social interaction'
Planter	• Beautification for home and the façade of the home block • Plants are indicators of seasonal change
Laundry	• Deeper and larger basin for handwashing clothes
Hooks near window	• To hang bird cages and air dry (preserve)foodstuffs
Access and circulation	• "Feng Shui" arrangement should be considered in Asian cultural environment (Chinese/ Japanese) • Barrier-free access
Railing	• Hand-rails installed in all circulation spaces and staircases
Rooftop and podium	• Rooftop and podium should be semi private and partially sheltered allowing different types of social activity to take place
Balcony	• Movement between interior and balcony should be level
Shape and size of balcony	• Balcony should be considered as an alternative social space for the residents
Balustrade	• Should not block the view
Lighting	• Natural lighting reminds people of natural temporal change • Two systems of artificial lighting to consider: overall lighting and bedside lamp / low level lighting
Emergency alarm	• Located in all main circulation corridors, common areas, places of high risk such as toilet, bathroom and kitchen
Switches and sockets	• Switches and sockets (preferably bigger in size) should be installed at hand reachable height (*e.g.* at waist level)
Entertainment device	• Home entertainment and internet access should be installed

Table 25.1. (c-d) Guidelines for designing a good living environment

(C) Home Unit	
Concept	**Guidelines**
Religious objects	• Estate management policy must observe the cultural and spiritual practices and habits of the residents
Artworks and souvenirs items	• Spaces for memorabilia and decoration display should be reserved
Ease-of-use factor	• Barrier-free principle should be respected
(D) Neighbourhood Space	
Concept	**Guidelines**
Neighbourhood environment	• Well designed neighbourhood environment respecting and supporting the everyday life patterns of older people
Activities nodes	• Integration of different types of places and activities to enrich the urban life
Strolling place	• Lively activities nodes and strolling places should be interlinked
Small parks	• Different kinds of shops surround the small park • Good balance between the "openness" and "closedness" of the space of the park
Large open space/District park	• District park within 20 minutes walk from an elderly residence
Building Edge	• Should be considered as a community gathering place , not just an abstract thin line in the plan demarcating the indoor and outdoor space
The "Façade" and "back" of buildings	• The 'Façade' faces an open space forming a pedestrian zone • The 'Back' faces vehicle and goods circulation routes
The Roads	• Materials with different texture for pavement construction can indicate changes from pedestrian walkway to vehicle circulation road • Speed control device to control the traffic
Streets/Pedestrian zone	• Should be vehicle-free with various facilities for social activities
Circulation route and seating spots	• Outdoor seating and resting places along circulation paths should be provided
Pedestrian facilities	• Pleasant promenade and seating near elderly residence

Table 25.1. (d-e) Guidelines for designing a good living environment

(D) Neighbourhood Space	
Concept	**Guidelines**
Outdoor lighting	• Creates a safe environment and encourages outdoor activities in the evening
Greenery and micro-climate	• Provides physical and psychological comfort
Linkages from interior to exterior	• The layering and sequence from private zone, community gathering zone and the neighbourhood should be well-balanced
Neighbourhood environment planning	• Social networking should aim at community building
(E) Management	
Concept	**Guidelines**
Management, designers and users	• Design direction and management policy should respect the everyday life culture of the residents • User participation management

25.4 References

Baltes MM, Carstensen LL (1996) The process of successful ageing. Ageing and Society, 16: 397–422

Bond J, Corner L (2004) Quality of life and older people. Open University Press, Maidenhead, UK

De Certeau M (1984) The practice of everyday life. University of California Press, Berkeley, CA, US

Kearns RA, Andrews GJ (2005) Placing ageing: positionings in the study of older people. In: Andrews GJ, Phillips DR (eds.) Ageing and place: perspectives, policy, practice. Routledge, London, UK

Lawton MP, Brody EM (1969) Assessment of older people: self-maintaining and instrumental activities of daily living. The Gerontologist, 9: 179–186

Norberg-Schulz C (1985) The concept of dwelling: On the way to figurative architecture. Electa, Milan, Italy

Wates N (ed.) (2000) The community planning handbook: how people can shape their cities, towns and villages in any part of the world. Earthscan, London, UK

Index of Contributors